I0502851

Shallow Groundwater in the Matanuska-Susitna Valley, Alaska— Conceptualization and Simulation of Flow

By Colin P. Kikuchi

Prepared in cooperation with the Alaska Department of Natural Resources

Scientific Investigations Report 2013–5049

U.S. Department of the Interior
U.S. Geological Survey

U.S. Department of the Interior
KEN SALAZAR, Secretary

U.S. Geological Survey
Suzette M. Kimball, Acting Director

U.S. Geological Survey, Reston, Virginia: 2013

Suggested citation:
Kikuchi, C.P., 2013, Shallow groundwater in the Matanuska-Susitna Valley, Alaska—Conceptualization and simulation of flow: U.S. Geological Survey Scientific Investigations Report 2013–5049, 84 p.

Contents

Contents

Figures

Figures

Tables

Conversion Factors and Datums

Conversion Factors

Multiply	By	To obtain
Length		
inch (in.)	2.54	centimeter (cm)
foot (ft)	0.3048	meter (m)
mile (mi)	1.609	kilometer (km)
Area		
acre	0.004047	square kilometer (km^2)
square mile (mi^2)	2.590	square kilometer (km^2)
Volume		
gallon (gal)	0.003785	cubic meter (m^3)
million gallons (Mgal)	3,785	cubic meter (m^3)
cubic foot (ft^3)	0.02832	cubic meter (m^3)
Flow rate		
acre-foot per year (acre-ft/yr)	1,233	cubic meter per year (m^3/yr)
foot per day (ft/d)	0.3048	meter per day (m/d)
gallon per day (gal/d)	0.003785	cubic meter per day (m^3/d)
cubic foot per second (ft^3/s)	0.02832	cubic meter per second (m^3/s)
inch per year (in/yr)	25.4	millimeter per year (mm/yr)
Specific capacity		
cubic foot per day per foot [(ft^3/d)/ft]	0.0929	cubic meters per day per meter [(m^3/d)/m]
gallon per minute per foot [(gal/min)/ft)]	0.2070	liter per second per meter [(L/s)/m]
Hydraulic conductivity		
foot per day (ft/d)	0.3048	meter per day (m/d)
Transmissivity*		
foot squared per day (ft^2/d)	0.09290	meter squared per day (m^2/d)

Temperature in degrees Celsius (°C) may be converted to degrees Fahrenheit (°F) as follows:

$$°F=(1.8×°C)+32.$$

Temperature in degrees Fahrenheit (°F) may be converted to degrees Celsius (°C) as follows:

$$°C=(°F-32)/1.8.$$

*Transmissivity: The standard unit for transmissivity is cubic foot per day per square foot times foot of aquifer thickness [(ft^3/d)/ft^2]ft. In this report, the mathematically reduced form, foot squared per day (ft^2/d), is used for convenience.

Specific conductance is given in microsiemens per centimeter at 25 degrees Celsius (µS/cm at 25 °C).

Concentrations of chemical constituents in water are given either in milligrams per liter (mg/L) or micrograms per liter (µg/L).

Datums

Vertical coordinate information is referenced to the North American Vertical Datum of 1988 (NAVD 88).

Horizontal coordinate information is referenced to the North American Datum of 1983 (NAD 83).

Altitude, as used in this report, refers to distance above the vertical datum.

Shallow Groundwater in the Matanuska-Susitna Valley, Alaska—Conceptualization and Simulation of Flow

By Colin P. Kikuchi

Abstract

The Matanuska-Susitna Valley is in the Upper Cook Inlet Basin and is currently undergoing rapid population growth outside of municipal water and sewer service areas. In response to concerns about the effects of increasing water use on future groundwater availability, a study was initiated between the Alaska Department of Natural Resources and the U.S. Geological Survey. The goals of the study were (1) to compile existing data and collect new data to support hydrogeologic conceptualization of the study area, and (2) to develop a groundwater flow model to simulate flow dynamics important at the regional scale. The purpose of the groundwater flow model is to provide a scientific framework for analysis of regional-scale groundwater availability.

To address the first study goal, subsurface lithologic data were compiled into a database and were used to construct a regional hydrogeologic framework model describing the extent and thickness of hydrogeologic units in the Matanuska-Susitna Valley. The hydrogeologic framework model synthesizes existing maps of surficial geology and conceptual geochronologies developed in the study area with the distribution of lithologies encountered in hundreds of boreholes. The geologic modeling package Geological Surveying and Investigation in Three Dimensions (GSI3D) was used to construct the hydrogeologic framework model. In addition to characterizing the hydrogeologic framework, major groundwater-budget components were quantified using several different techniques. A land-surface model known as the Deep Percolation Model was used to estimate in-place groundwater recharge across the study area. This model incorporates data on topography, soils, vegetation, and climate. Model-simulated surface runoff was consistent with observed streamflow at U.S. Geological Survey streamgages. Groundwater withdrawals were estimated on the basis of records from major water suppliers during 2004-2010. Fluxes between groundwater and surface water were estimated during field investigations on several small streams.

Regional groundwater flow patterns were characterized by synthesizing previous water-table maps with a synoptic water-level measurement conducted during 2009. Time-series water-level data were collected at groundwater and lake monitoring stations over the study period (2009–present). Comparison of historical groundwater-level records with time-series groundwater-level data collected during this study showed similar patterns in groundwater-level fluctuation in response to precipitation. Groundwater-age data collected during previous studies show that water moves quickly through the groundwater system, suggesting that the system responds quickly to changes in climate forcing. Similarly, the groundwater system quickly returns to long-term average conditions following variability due to seasonal or interannual changes in precipitation. These analyses indicate that the groundwater system is in a state of dynamic equilibrium, characterized by water-level fluctuation about a constant average state, with no long-term trends in aquifer-system storage.

To address the second study goal, a steady-state groundwater flow model was developed to simulate regional groundwater flow patterns. The groundwater flow model was bounded by physically meaningful hydrologic features, and appropriate internal model boundaries were specified on the basis of conceptualization of the groundwater system resulting in a three-layer model. Calibration data included 173 water-level measurements and 18 measurements of streamflow gains and losses along small streams.

Comparison of simulated and observed heads and flows showed that the model accurately simulates important regional characteristics of the groundwater flow system. This model is therefore appropriate for studying regional-scale groundwater availability. Mismatch between model-simulated and observed hydrologic quantities is likely because of the coarse grid size of the model and seasonal transient effects. Next steps towards model refinement include the development of a transient groundwater flow model that is suitable for analysis of seasonal variability in hydraulic heads and flows. In addition, several important groundwater budget components remain poorly quantified—including groundwater outflow to the Matanuska River, Little Susitna River, and Knik Arm.

Introduction

The Matanuska-Susitna Valley aquifer system supplies water to more than 50,000 residents. Population growth has been heavily concentrated in the cities of Palmer, Wasilla, and Houston and in the Matanuska-Susitna Borough (MSB) core area between Palmer and Wasilla. Increasing residential and industrial development in this area over the last 20 years has led to concerns regarding the long-term availability of groundwater in adequate quantities to meet the demands of the population. The population residing on lots outside the service area for municipal water and city sewer in Palmer and Wasilla relies on small-capacity domestic wells for water supply and septic systems for wastewater disposal. As a result, concerns have also been raised about the susceptibility of shallow groundwater and surface water to contamination from septic leachate. Groundwater availability and quality will become increasingly important under current and future stresses from population growth, and management of the groundwater resources in this area is limited by the lack of information about the aquifer system in the core area. A cooperative study between the Alaska Department of Natural Resources (ADNR) and the U.S. Geological Survey (USGS) was initiated in 2009 to address these issues.

The cooperative study documented in this report included two main objectives that were developed concurrently. The first objective was to compile existing data and collect new data supporting detailed characterization of hydrogeologic conditions in the regional aquifer system. The second objective was to develop a numerical model simulating the groundwater flow system. The analysis of new and existing data identified important hydrogeologic features and flow conditions deemed important for a regional-scale groundwater flow model. At the same time, computer models of increasing complexity were used throughout the study period to guide data collection and identify data that would contribute to a conceptual understanding of the groundwater flow system.

Purpose and Scope

This report documents the development of conceptual and numerical models of groundwater flow for the Matanuska-Susitna Valley aquifer system, utilizing existing and new data sources. The conceptual model is based on aquifer extent, thickness, and hydraulic properties, combined with analysis of field data pertaining to aquifer inflows and outflows and the distribution of water levels. The conceptual model provides the basis for the numerical model, which was evaluated against existing hydrologic data. The numerical model may be used to improve understanding of groundwater hydrology in the study area. The scale of this model limits its use to regional-scale analysis; however, the numerical model was developed in a way to allow for future compatibility with local-scale hydrologic models.

Description of Study Area

The Matanuska-Susitna Valley (fig. 1) is in Upper Cook Inlet, south-central Alaska, approximately 60 mi north of Anchorage. Much of the population resides outside the incorporated cities of Palmer, Wasilla, and Houston. In particular, population density is relatively high in the MSB core area, between the cities of Palmer and Wasilla. The Matanuska-Susitna Valley is within the Cook Inlet aquifer system described by Glass (2001) and Brabets and others (1999). The study area is bounded by two glacial rivers: the Matanuska River to the east and the Susitna River to the west. A third major river, the Little Susitna River, flows out of and along the base of the Talkeetna Mountains, before turning south past the city of Houston and flowing through the Susitna Lowlands into Cook Inlet. The Susitna River is nearly 40 mi west of the core area, and the study area was extended to this feature because of its role as a regional boundary for shallow groundwater flow in the study area. Spatially extensive alluvial aquifers are present beneath and surrounding these three rivers; however, only the alluvial aquifers underlying the Matanuska River and Little Susitna River are of importance for water supply in the core area. Other productive aquifers in this area include unconsolidated sediments of glacial and glaciofluvial origin. The thickness of the unconsolidated sediments increases from approximately 50 to over 500 ft, moving south from the Talkeetna Mountains into Knik Arm. Groundwater is present under confined and unconfined conditions in the unconsolidated sediments and is also present in sedimentary bedrock underlying the unconsolidated sediments.

The Matanuska-Susitna Valley is in a transition zone between maritime and continental climates; average air temperatures in the study area range from 10°F during the winter to 70°F during the summer. The average annual total precipitation measured at the Matanuska Agricultural Experimental Station (MAES) near Palmer from 1917 to 2010 is 15.3 in. Precipitation typically increases with elevation; the average annual total precipitation measured at the Independence Mine Natural Resource Conservation Service (NRCS) Snowpack Telemetry (SNOTEL) station near Hatcher Pass is 36.4 in. Hydrologic processes at the land surface are strongly influenced by snow and ice cover which typically last from October until April. The dominant vegetation type in the study area varies along an east-west transect, transitioning from deciduous forest and agricultural fields between the Matanuska River and Wasilla to woody and emergent wetlands in the area surrounding Big Lake.

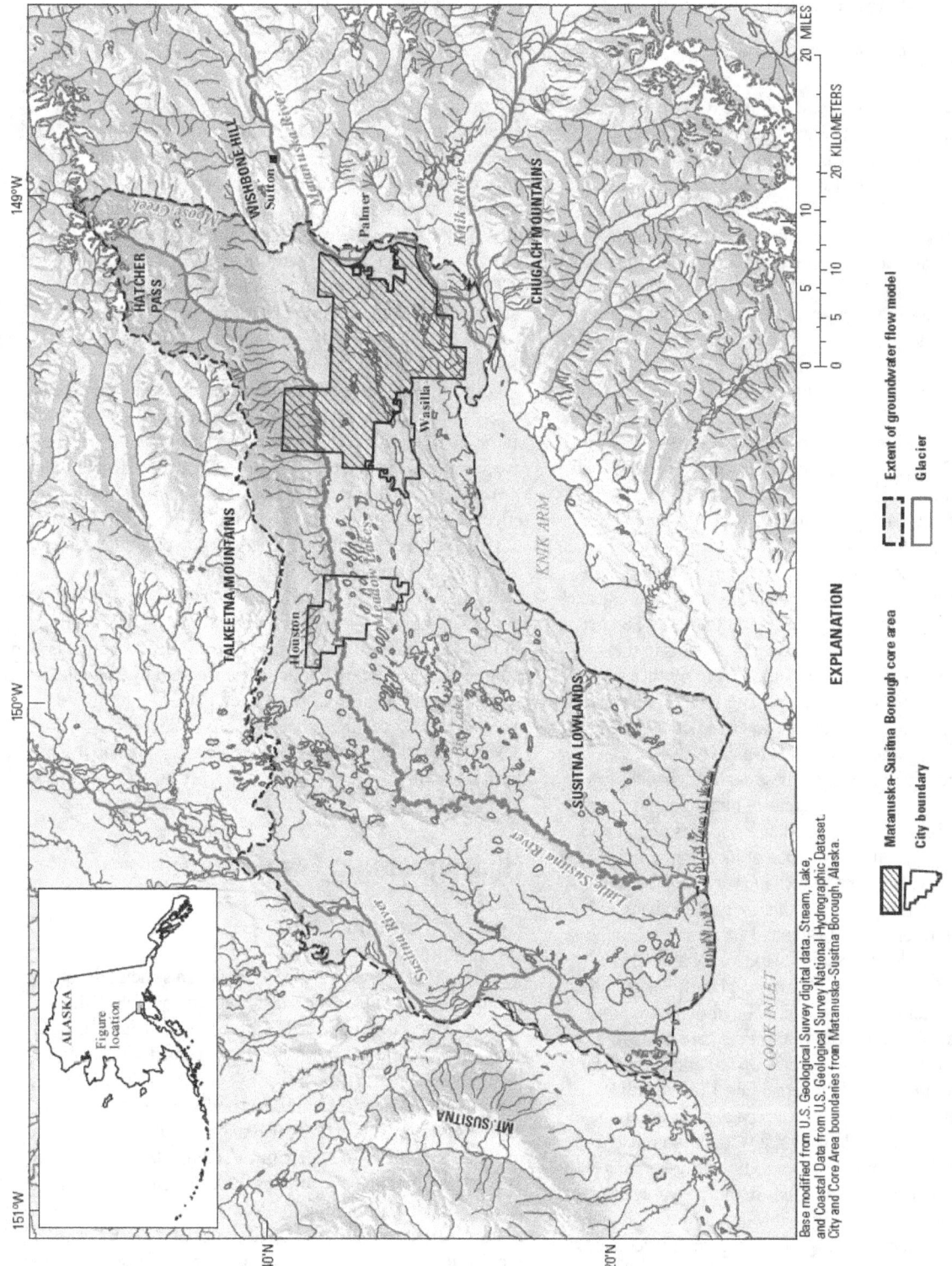

Base modified from U.S. Geological Survey digital data. Stream, Lake, and Coastal Data from U.S. Geological Survey National Hydrographic Dataset. City and Core Area boundaries from Matanuska-Susitna Borough, Alaska.

EXPLANATION

Matanuska-Susitna Borough core area

City boundary

Extent of groundwater flow model

Glacier

Figure 1.   Location of study area, Matanuska-Susitna Valley, Alaska, and extent of groundwater flow model.

Previous Investigations

Trainer (1960) provides the earliest description of hydrogeology in the study area, focusing primarily on groundwater conditions in the Matanuska River alluvial aquifer. This aquifer was an important source of water following agricultural colonization of the area near Palmer beginning in the 1930s. His study included detailed descriptions of unconsolidated sediments in the study area, including analysis of grain size and estimation of aquifer hydraulic properties from aquifer tests. From analysis of groundwater hydrographs of five wells, Trainer inferred that in-place groundwater recharge takes place primarily following late summer rains.

Building on descriptions of Quaternary geology and glacial-to-interglacial history (Freethey and Sculley, 1980; Reger and Updike, 1983), Jokela and others (1990) developed a conceptual model of the hydrogeologic controls on groundwater flow in the Matanuska-Susitna Valley and used lithologic information and water levels from 3,600 water wells to construct hydrogeologic sections in the area between Palmer and Big Lake. Water levels were compiled to generate a contour map of the water-table altitude in the core area. The water-table map indicates flow through the groundwater system is driven primarily by recharge in the Little Susitna River Valley, following north to south flow paths in the area surrounding Wasilla and transitioning to northeast to southwest flow paths in the area near Big Lake. Jokela and others (1990) also distinguished between regional-scale and local-scale groundwater flow paths, which in some cases are oriented in opposite directions. These scale-dependent flow paths occur in wetlands and in areas of hummocky terrain underlain by glacial moraine and kame deposits and likely result from the controlling influence of local topography when the water table is very close to the land surface. Jokela and others (1990) used the results of this analysis to characterize interactions between groundwater and surface water in the principal watersheds of the study area. Jokela and others (1990) also used the water-table map to classify lakes in the study area according to their dominant hydrologic regime, estimating that 351 of the 439 lakes in the core area are seepage lakes with no inlet or outlet, 56 are drainage lakes with inlets and outlets, and the remaining 32 have either inlets or outlets only. Many of the seepage lakes are in the Meadow Lakes area, in relatively flat wetland-dominated terrain. Wetlands mapping and delineation efforts in this area (Gracz, 2009) indicate that wetlands in this area are sustained by seasonal precipitation and shallow groundwater discharge. The relation between shallow groundwater in the wetlands surrounding Big Lake and the regional groundwater flow system has not been clearly established.

Moran and Solin (2006) compiled water levels in nearly 800 wells penetrating unconfined aquifers and used those levels to construct a contour map of water-table altitude in the core area. The configuration of the water table was similar to that described by Jokela and others (1990). On the basis of the apparent water-table surface and similarities and the geochemical and isotopic composition of surface water and groundwater, Moran and Solin (2006) also identified locations of potentially substantial groundwater/surface-water interaction. Geochemical and isotopic data for groundwater throughout Upper Cook Inlet Basin were also collected and compiled by Glass (2001). Of particular importance from this study were groundwater apparent ages calculated from tritium:helium-3 (^{3}H:^{3}He) ratios and chlorofluorocarbons (CFCs). These analyses of groundwater samples collected from confined and unconfined aquifers in the Matanuska-Susitna Valley showed the groundwater is quite young (less than 60 years from time of recharge).

From the investigations discussed above and the public records of borehole lithology and groundwater geochemistry in the study area, there is some consensus on the conceptual aspects of groundwater flow in the Matanuska-Susitna Valley. However, only one quantitative assessment of pumping effects on groundwater resources has been published. Munter (2010) examined potential impacts of gravel-pit development in the unconfined alluvial aquifer south of Palmer using the Theis equation for drawdown near a pumping well, making the assumption that gravel-pit dewatering is hydraulically similar to groundwater pumping. Using independently derived estimates of aquifer hydraulic properties and hydrogeologic sections constructed from water-well data, Munter found that water-level declines of several feet associated with the development of a gravel pit could extend as much as 1.5 mi from the site of the gravel pit.

Hydrogeologic Framework

Geologic Setting

The Matanuska-Susitna Valley is in a structural trough bounded by two faults: the Castle Mountain Fault to the north and the Border Ranges Fault to the southeast (fig. 2). A simplified geologic history of the Matanuska-Susitna Valley includes three major time periods. The Mesozoic Era geology includes the roots of the Talkeetna and Chugach Mountains, comprising intrusive and extrusive igneous rocks adjacent to metamorphic rocks and deep, thick sedimentary layers. The Tertiary Period geology includes the Chickaloon and Arkose Ridge Formations abutting the Talkeetna Mountains, as well as the oil-bearing Kenai Group in the Cook Inlet Basin. Finally, the basement rock generated during these two periods is overlain by unconsolidated Quaternary sediments of glacial and nonglacial origin. These sediments are highly permeable in places and comprise the most productive aquifers in the Matanuska-Susitna Valley.

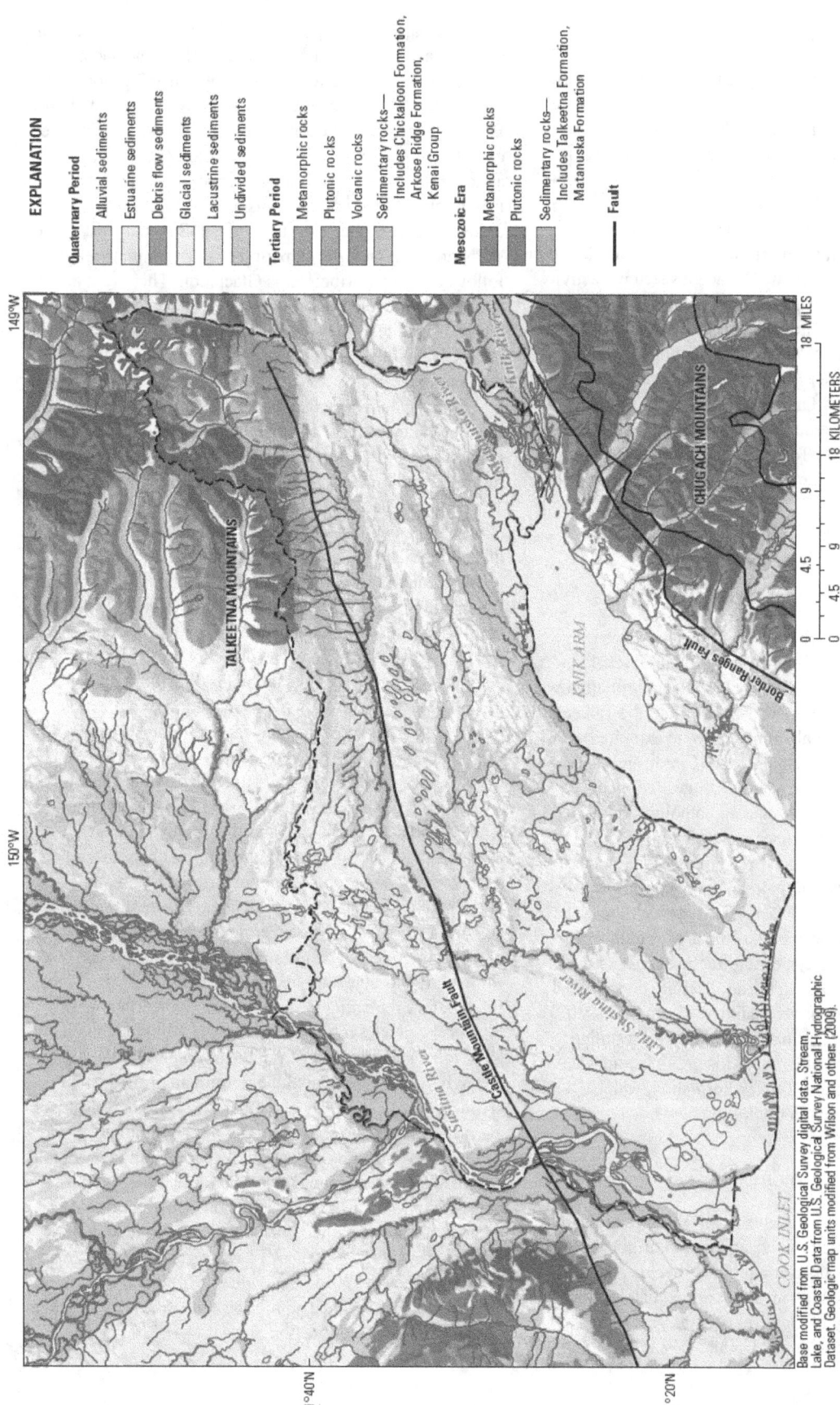

EXPLANATION

Quaternary Period

Alluvial sediments

Estuarine sediments

Debris flow sediments

Glacial sediments

Lacustrine sediments

Undivided sediments

Tertiary Period

Metamorphic rocks

Plutonic rocks

Volcanic rocks

Sedimentary rocks—
Includes Chickaloon Formation,
Arkose Ridge Formation,
Kenai Group

Mesozoic Era

Metamorphic rocks

Plutonic rocks

Sedimentary rocks—
Includes Talkeetna Formation,
Matanuska Formation

Fault

Base modified from U.S. Geological Survey digital data. Stream,
Lake, and Coastal Data from U.S. Geological Survey National Hydrographic
Dataset. Geologic map units modified from Wilson and others (2009).

Figure 2.   Matanuska-Susitna Valley, southcentral Alaska.

Mesozoic Era Geology

The Matanuska-Susitna Valley is bounded to the south by the Border Ranges fault. This fault separates plate segments of different origins: the older Peninsular terrane northwest of the fault and the Chugach terrane southeast of the fault. The Peninsular terrane—primarily intrusive and extrusive igneous rocks—was generated by late Triassic to middle Jurassic island arc magmatism (Burns, 1985; Trop and Ridgway, 1999); these rocks are on the north side of the Castle Mountain Fault, east of the study area. Late Cretaceous tectonic activity resulted in accretionary wedges of clastic/volcaniclastic and marine sedimentary rocks, now known, respectively, as the Talkeetna Formation and Matanuska Formation (Kirschner and Lyon, 1973). The latter is at least 10,000 ft thick and is presently exposed in the Matanuska Valley east of Sutton. Non-marine and estuarine deposition contributed the bulk of sediments during the early Tertiary depositional cycle; arc magmatism and accretion continued into the Paleocene and Eocene depositional cycles. The latter events are more directly relevant to the study area.

Tertiary Period Geology

The Paleocene and Eocene depositional cycle comprises non-marine sediments that lithified to form the Chickaloon and Arkose Ridge Formations. The Chickaloon Formation was likely derived from metamorphic rocks north of the Talkeetna Mountains and contains coal seams that were exploited through the late 1960s at Wishbone Hill (Clardy and others, 1984). The Arkose Ridge Formation includes sediments derived from the Alaskan and Aleutian batholiths, deposited farther north in the trough (Kirschner and Lyon, 1973). Both formations are presently exposed in the front range of the Talkeetna Mountains and underlie the unconsolidated Quaternary sediments of the Matanuska-Susitna Valley. Along the northern boundary of the core area, a number of supply wells penetrate water-bearing fracture zones of the Chickaloon Formation. During the late Tertiary depositional cycle, estuarine and non-marine sediments were deposited deeper in the Cook Inlet Basin structural trough, resulting in the 25,000 ft thick Kenai Group. This group includes the Hemlock Conglomerate and Tyonek Formation, which contain the majority of oil deposits in Cook Inlet.

Quaternary Period Geology

The unconsolidated sediments in the Matanuska-Susitna Valley are of glacial, glaciofluvial, and fluvial origin and include many highly productive aquifers. There is general agreement on the chronology up until approximately 15–25 ka. The summary of Quaternary history presented below is adapted from Reger and Updike (1983).

The earliest glaciation in the Upper Cook Inlet for which substantial evidence exists was the Mt. Susitna glaciation. This event is assigned to late Pliocene–early Pleistocene time on the basis of relative age criteria and correlation with the Browne Glaciation in the north-central Alaska Range. The extent of this glaciation is not well established; however, a maximum elevation of 4,000 ft above mean sea level is tentatively assigned on the basis of ice-scoured surfaces and old erratics on higher elevations of Mt. Susitna, west of the study area. An interglacial period of indeterminate length was followed by the Caribou Hills Glaciation. This event also is assigned to the late Pliocene–early Pleistocene time using ratios of granite to greywacke-argillite erratics. The Caribou Hills Glaciation extended down the length of Cook Inlet into the Pacific Ocean.

Late Quaternary glaciations covering the Matanuska-Susitna Valley are better constrained than early Quaternary glaciations with respect to age and spatial extent due to improved radiometric, isotopic, and stratigraphic evidence. The Eklutna Glaciation is estimated to have lasted from 200 to 130 ka and was of lesser extent than the two previous glaciations. The top of the ice is estimated at 2,350–3,100 ft above mean sea level and covered the Susitna Lowlands and Cook Inlet. Following an interglacial period of approximately 55,000 years, the Knik Glaciation (75–50 ka) extended partway down Cook Inlet. The elevation of the ice was again lower during this glaciation (700–2,400 ft above mean sea level). The Naptowne Glaciation was the most recent glaciation and deposited nearly all the surficial glacial sediments found in the Matnuska-Susitna Valley today. This event is subdivided into two distinct glacial advances.

During the first advance (145–35 ka), the bilobate Knik-Matanuska glacier advanced from the Knik and Matanuska River Valleys into Cook Inlet, likely extending as far south as Anchorage, but not past the Susitna lowlands in the southwest part of the study area. During the second advance (30–10 ka), the glacier readvanced into the waters of Cook Inlet, leaving a prominent end moraine—the Elmendorf Moraine—across a broad swath of Upper Cook Inlet, before retreating to the northwest. The Matanuska lobe of the glacier retreated much more quickly than the Knik lobe, and meltwater from the Matanuska lobe flowed through the stagnant Knik lobe to form an array of esker and crevasse features (Clardy and others, 1984). These features are prominent in the area south of Palmer and likely exert control over local groundwater flow patterns in their vicinity. At this time, the stagnant Knik lobe formed the southern bank of the Matanuska River, directing the river to flow in a west-southwest direction. The paleochannel deposits associated with the Matanuska River are now identified as outwash deposits (Wilson and others, 2009) and constitute an important unconfined aquifer bisecting the core area in a northeast-southwest direction.

The Meadow Lakes area west of Wasilla is characterized by undulating terrain with lakes occupying topographic depressions. Several contradictory interpretations of these surficial features have been proposed. Trainer (1960) attributed these features to supraglacial ablation till deposited at regular intervals during the retreat of the Matanuska Lobe. Reger and Updike (1983) interpreted these features as Rogen moraines, periodically deposited as terminal moraines with seasonal advance and retreat cycles of the Matanuska Lobe. More recently, Wiedmer and others (2010) postulated that the undulating terrain originated during a catastrophic outburst flood from an ancient glacial lake during the stagnation or retreat of the Matanuska Lobe. The first two interpretations have relatively similar implications with respect to hydrogeologic characterization, in that the alternating glacial and meltwater deposits should explain lithologic variability along an east-west transect. According to the third interpretation, the surficial deposits in this area would constitute a relatively uniform hydrogeologic unit. Stratigraphic interpretation and hydrogeologic characterization in the Meadow Lakes area depends on the choice of interpretation; for this report, the interpretation of Reger and Updike (1983) was adopted.

Aquifer Extent and Thickness

Data Sources

For the purposes of this study, the most important geologic data in the study area were well driller's logs qualitatively describing the lithology at regular depth intervals. To assess the distribution of geologic deposits in the shallow subsurface, driller's logs from 327 water wells within the study area were obtained from the Alaska Well Log Tracking System (WELTS) database (Alaska Department of Natural Resources, 2009). The mean and median depths of the selected wells were 161 and 138 ft, respectively. In addition to driller's logs from WELTS, 10 oil and gas exploration mud logs were obtained from the Alaska Oil and Gas Conservation Commission (Alaska Department of Administration, 2010). For these exploration wells, the mean and median depths were 1,745 and 1,525 ft, respectively. The wells used in this part of the study span the range of the core area, with at least one well per square mile (fig. 3). Some of the driller's logs reported latitude and longitude of the well; others included only township-range section (TRS) aliquot identification. For the latter, the geographic location was estimated using TRS maps. For each well inventoried, the well location and lithologic data in each driller's log, including qualitative description of geologic material encountered at different depths, were extracted from the driller's logs and stored in a database (appendix A).

Many drillers' logs used in this study show the depth at which bedrock was encountered and were used to identify the interface between consolidated and unconsolidated sediments. The base of the Tertiary rock south of the Castle Mountain Fault was estimated by Shellenbaum and others (2010) on the basis of seismic profiles and oil and gas exploration wells. These results were used to define the base of the Tertiary rock hydrogeologic unit in the geologic sections. In some cases, sections were extended far beyond the boundaries of the study area for the purposes of geologic model calculation. Outside the core area—especially on the western shore of the Susitna River and north of the Little Susitna River, on the flanks of the Talkeetna Mountains—fewer driller's logs were available. In these areas of sparse data, geologic control points were constructed and used to ensure that boundaries with unit surfaces were consistent with regional structural and depositional trends.

Once all borehole data had been compiled, lithology records were imported into a computer program, Geological Surveying and Investigation in Three Dimensions (GSI3D), documented by Mathers and others (2011). GSI3D is a platform for merging borehole lithologic data with geologic maps and digital elevation models (DEM). GSI3D supports geologic interpretation and construction of two-dimensional geologic sections and calculation of three-dimensional subsurface geometry. In this study, GSI3D was used to construct the hydrogeologic framework model, describing the extent and thickness of hydrogeologic units in the study area. Constructing the hydrogeologic framework model entailed (1) developing a lithologic-unit classification scheme, (2) defining appropriate hydrogeologic units, (3) constructing and manually correlating hydrogeologic sections with geologic maps, (4) estimating boundaries that define the lateral extent of hydrogeologic units, and (5) calculating the extent and thickness for each hydrogeologic unit.

Lithologic Units

Prior to the creation of geologic sections, a lithologic classification scheme was developed. Thirty-seven distinct lithologic descriptors were identified from the driller's logs used in this study. Each lithologic descriptor was grouped into one of five lithologic units: coarse unconsolidated sediments, fine unconsolidated sediments, heterogeneous sediments or diamict, sedimentary rock, and crystalline or metamorphic rock (table 1). The first three units include homogeneous and heterogeneous lithologic descriptors. For example, both gravel and sandy gravel were classified as coarse unconsolidated sediments under this scheme. Lithologic descriptors including a mixture of coarse sediments and fine sediments —for example, silt, gravel, and clay—were classified as heterogeneous sediments or diamict. Many driller's logs report the presence of hardpan and glacial till; these lithologies were similarly classified as heterogeneous sediments or diamict.

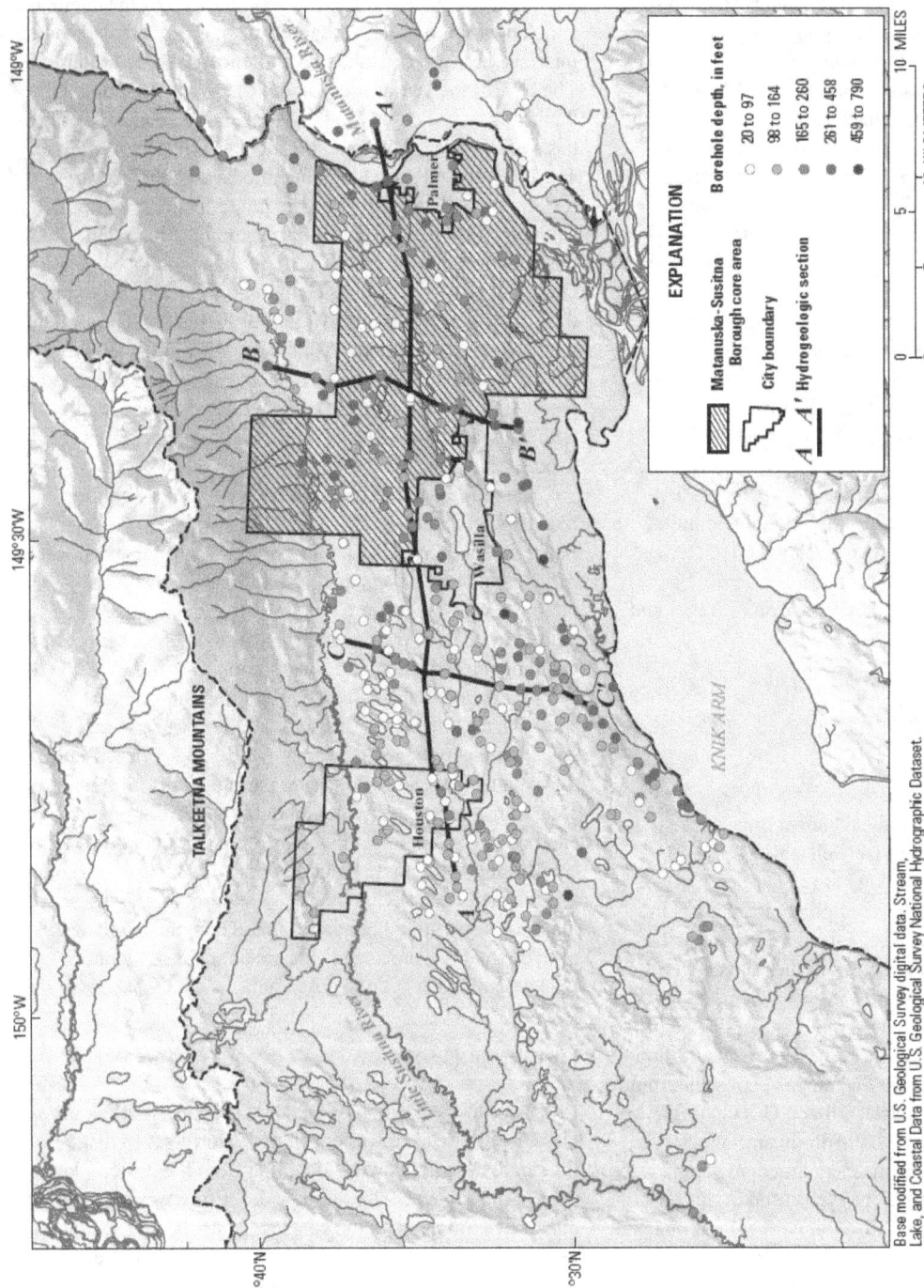

Base modified from U.S. Geological Survey digital data. Stream,
Lake, and Coastal Data from U.S. Geological Survey National Hydrographic Dataset.
Geologic map units modified from Wilson and others (2009).

Figure 3.   Location of boreholes with lithologic data and representative hydrogeologic sections, Matanuska-Susitna Valley, Alaska.

Table 1. Lithologic unit classification scheme used in hydrogeologic framework model, Matanuska-Susitna Valley, Alaska.

Lithologic descriptor	Lithologic unit code
Coarse unconsolidated sediments	
Overburden/soil	Coarse
Sand	Coarse
Gravel	Coarse
Gravel, boulders	Coarse
Cobble	Coarse
Boulders	Coarse
Sand and gravel	Coarse
Gravel and sand	Coarse
Silt and sand	Coarse
Sand and silt	Coarse
Sand and clay	Coarse
Sand, silt, and gravel	Coarse
Silt, sand, and gravel	Coarse
Gravel and silt	Coarse
Sand, gravel, clay	Coarse
Boulder, gravel, clay	Coarse
Crystalline/metamorphic rock	
Crystalline rocks	Mesorx
Greenstone	Mesorx
Andesite	Mesorx
Fine unconsolidated sediments	
Silt	Fine
Clay	Fine
Clay and sand	Fine
Silt and gravel	Fine
Silt and clay	Fine
Clay and gravel	Fine
Gravel and clay	Fine
Silt and cobbles	Fine
Heterogeneous sediments/diamict	
Till/hardpan	Diamict
Silt, gravel, clay	Diamict
Sedimentary rock	
Sandstone	Tertrx
Limestone	Tertrx
Shale	Tertrx
Coal	Tertrx
Siltstone	Tertrx
Claystone	Tertrx
Conglomerate	Tertrx

Shallow boreholes drilled near the Talkeetna Mountain front, as well as deeper oil and gas exploration boreholes, penetrate sequences of sedimentary rocks, including sandstone, siltstone, claystone, shale, and coal. These lithologic descriptors were classified as sedimentary rock. Finally, crystalline and metamorphic rocks were encountered in several wells east of Palmer and likely correspond to exposures of those rocks in the Chugach Mountains and in the Matanuska River Valley.

Hydrogeologic Units

The boundaries of hydrogeologic units were determined using the conceptual understanding of Quaternary depositional environments and the spatial distribution of lithologic units. Quaternary depositional environments ranged from high-energy—including debris flows, alluvium, and glacial outwash—to low-energy—including estuarine and lacustrine deposition. In general, textural characteristics of the sediment are correlated with this range of depositional environments. For example, outwash and alluvial deposits generally include larger particle sizes such as gravel and cobbles. Similarly, estuarine and lacustrine deposits generally include smaller particle sizes such as silt and clay. Unconsolidated sediments comprising large particle sizes are typically more permeable than those comprising small particle sizes, and well-sorted deposits tend to be more permeable than poorly sorted deposits. Therefore, conceptual understanding of depositional environments is useful in defining and identifying relevant hydrogeologic units. For this study, three distinct depositional environments—listed in geochronological order—were identified during the Quaternary Period: (1) glacial transport and deposition, (2) lacustrine, estuarine, and glacioestuarine deposition, and (3) glacial outwash and more recent alluvium. These depositional environments correspond to three hydrogeologic units defined in this study, comprising recent unconsolidated sediments exposed at the land surface: (1) Naptowne Moraine, (2) Fine Sediments, and (3) Holocene Outwash and Alluvium. The Fine Sediments hydrogeologic unit spans the entire study area and includes areas of active estuarine deposition along Knik Arm, south of Palmer, as well as areas of Pleistocene sand deposits west of the Little Susitna River. The Pleistocene glacioestuarine sediments west of the Little Susitna River primarily comprise well-bedded and sorted sand (Wilson and others, 2009) and likely are more permeable than the recent estuarine sediments along Knik Arm, south of Palmer. Surficial Quaternary unconsolidated sediments from figure 2 were classified in one of these three depositional environments (table 2). In addition to these three

Table 2. Age, depositional environment, and lithologic units for each hydrogeologic unit in the Matanuska-Susitna Valley, Alaska.

[Abbreviation: –, not present at land surface]

Geologic age	Geologic units (map unit)	Depositional environment	Hydrogeologic unit	Primary lithologic unit	Secondary lithologic unit
Quaternary period	Alluvial sediments	Main and side channels of glacial outwash and modern rivers streams	Holocene Outwash and Alluvium	Coarse	Fine
Quaternary period	Estuarine sediments	In and surrounding tide zones, tidal channels, in the ocean near glaciers	Fine Sediments	Fine	Coarse
Quaternary period	Glacial sediments, lacustrine sediments	Areas in and surrounding glaciated zones, glacial meltwater ponds and lakes	Naptowne Moraine	Poorly sorted/ diamict	Fine, coarse
Quaternary period Tertiary period	–	Numerous depositional environments	Lower Permeable Sediments	Coarse	Fine
Tertiary period	Tertiary sedimentary rocks	Numerous depositional environments	Tertiary Sedimentary Rocks	Sedimentary rock	–
Mesozoic era	Plutonic, volcanic, metamorphic rocks, mesozoic sedimentary rocks	Numerous depositional environments	Bedrock	Bedrock	–

hydrogeologic units, a fourth hydrogeologic unit—(4) Lower Permeable Sediments—was used to describe unconsolidated sediments present at depths too great to be associated with the Naptowne glaciation. Lower Permeable Sediments are most commonly found at depth beneath diamict deposits beginning near the city of Wasilla and moving south-southwest away from the Talkeetna Mountains. These permeable sediments are commonly water bearing and are exploited for groundwater; therefore, these permeable sediments are hydrogeologically significant. However, geologic evidence is insufficient for detailed description of these sediments or their corresponding depositional environment. Therefore, they are simply referred to as Lower Permeable Sediments. Unconsolidated sediments in the study area are underlain by rocks of Tertiary and Mesozoic age. Tertiary sedimentary rocks—in particular, terrestrially derived sandstone-shale-siltstone-coal sequences of the Chickaloon and Arkose Ridge Formations–are more permeable than Mesozoic-age rocks and constitute a semi-productive aquifer in the study area. These rocks are exposed at the land surface along the front of the Talkeetna Mountains and are encountered beneath unconsolidated sediments throughout the area west of Palmer. A fifth hydrogeologic unit—Tertiary Sedimentary Rocks— was defined to include these rocks. To the east and southeast of Palmer, Mesozoic Era rocks are encountered beneath the

unconsolidated sediments and are exposed along the front of the Chugach Mountains. These include less-permeable crystalline and metamorphic rocks, as well as marine-derived sedimentary rocks. West of Palmer, these Mesozoic Era rocks underlie sedimentary rocks of the Tertiary Period. In the hydrogeologic framework for this study, Mesozoic Era rocks are treated as impermeable basement rock.

The four hydrogeologic units comprising the unconsolidated sediments are very broadly defined and encompass substantial variability in sediment texture. For example, a single glacial outwash deposit could include a blend of fine sediments such as clay and silt and coarse sediments such as gravel and boulders. Furthermore, sediment lenses commonly were observed within each hydrogeologic unit. Taking into consideration the heterogeneity internal to each hydrogeologic unit, lateral interpolation between boreholes penetrating heterogeneous deposits or lenses could lead to error in hydrogeologic interpretation. Therefore, each hydrogeologic unit was defined to be internally heterogeneous. This scheme allowed some lenses of geologic material— texturally different from surrounding material—to be included in the same hydrogeologic unit. This approach was appropriate on the basis of the processes underlying glacial deposition, the borehole data, and the regional scale of the hydrogeologic framework.

Hydrogeologic Sections

The spatial distribution of lithologic units was interpreted within the framework of the hydrogeologic units described above and was used to construct 28 geologic sections. Representative sections are displayed in figures 4A-C. The sections show the contact between different hydrogeologic units. All three sections, for example, show how the Holocene Outwash and Alluvium units generally are located within incisions into either the Naptowne Moraine or the Tertiary Sedimentary Rocks and are generally associated with current surface-water features. The base of the Tertiary Sedimentary Rocks identified by Shellenbaum and others (2010) is shown in figure 4A. The procedure for constructing sections was as follows. First, each surficial deposit included on the geologic map of Wilson and others (2009) was reclassified into one of the exposed hydrogeologic deposits listed above. For each section, the locations of coarse sediments associated with Quaternary alluvium and glacial outwash deposits were identified from the surficial geologic map, and these features were manually drawn into the sections. Next, sequentially lower hydrogeologic units were identified on the basis of borehole lithologic units. For the boreholes used to construct sections A-A', B-B', and C-C' in figures 4A-C, the sequences of lithologic units encountered are listed in appendix A. Many of the borehole lithologies used in this analysis exhibited vertical heterogeneity.

A common case of vertical lithologic heterogeneity within a single borehole is discussed to provide an example of how such heterogeneity is accounted for in the hydrogeologic sections. Well-sorted coarse and fine sediments were commonly encountered between layers of diamicton within a single borehole. In some cases, such lithologic variability was interpreted as interfingering between different hydrogeologic units—for example, the variable extent of Lower Permeable Sediments in contact with Naptowne Moraine deposits. Otherwise, lithologic variability within a single borehole was interpreted as different facies within a single hydrogeologic unit. The latter interpretation includes well-sorted sediments within the Naptowne Moraine occurring as kame or esker deposits, as well as lenses formed from periodic retreats and readvances of the ice. Under this latter interpretation, the geologic data are insufficient to map the extent of such intra-moraine-sorted deposits, and the deposits were, therefore, grouped together with the Naptowne Moraine deposits.

Extent of Hydrogeologic Units

Surficial geologic maps were used in conjunction with the hydrogeologic sections to draw envelopes encompassing the lateral extent of each hydrogeologic unit. The Lower Permeable Sediments are not exposed at the land surface, so the lateral extent of this unit was determined using the hydrogeologic sections alone. Once the lateral extent of each unit had been defined, the three-dimensional hydrogeologic framework model was calculated in GSI3D.

The hydrogeologic framework model includes the elevations of the base of hydrogeologic units, including the Tertiary Sedimentary Rocks, Lower Permeable Sediments, Naptowne Moraine, Fine Sediments, and Holocene Outwash and Alluvium. Contour maps for the bases of each hydrogeologic unit are shown in figures 5A-E.

Aquifer Hydraulic Properties

The productivity of aquifers in the Matanuska-Susitna Valley is highly variable. Some wells in highly permeable formations withdraw groundwater at several hundred gallons per minute with very little drawdown. For other wells in less-permeable formations, large drawdowns are observed even when the screened or open interval of the well is very long. In general, the most productive wells are those finished in the alluvial terrace of the Matanuska River near Palmer (Holocene Outwash and Alluvium hydrogeologic unit) and paleochannel deposits of the Matanuska River running through the city of Wasilla. Aquifer productivity and response of the groundwater system to stresses depend on hydraulic properties such as transmissivity, specific yield, and storativity. Estimates of aquifer transmissivity are available from several aquifer tests performed within the core area; however, the data from these tests were not analyzed to estimate storativity and specific yield. Additionally, values of specific capacity were estimated from well flow tests performed on a number of community supply wells.

Aquifer-Test and Slug-Test Data

Several aquifer tests were performed at a gravel mining site southwest of Palmer, including single-well aquifer tests performed on two wells and one test performed in a large pond (D. Brailey, Brailey Hydrologic Consultants, written commun., 2011). Lithologic information was not available for the two wells, but both were thought to be finished in a gravel formation. Analysis of recovery data for the single-well aquifer tests yielded estimates of hydraulic conductivity of 1,400 and 53 ft/d. Analysis of drawdown and recovery data from an aquifer test on the large pond yielded an estimated hydraulic conductivity of 1,400 ft/d. These values are typical of well-sorted sand and gravel formations.

Numerous tests of aquifer hydraulic properties have been performed at a proposed mine site near the confluence of Moose Creek with the Matanuska River (Usibelli Coal Mine, Inc., 2009). Although this area is outside the study area, the geologic units in this area are similar to those within the study area. Aquifer tests performed on wells screened at multiple intervals in the coal-bearing Wishbone Hill Formation yielded estimates of hydraulic conductivity of 0.01 and 0.006 ft/d. Slug-test data indicate a range of values for different geologic materials. Values of hydraulic conductivity varied from 0.005 to 4.7 ft/d in bedrock formations, from 10.3 to 52.4 ft/d in stream alluvium, and from 0.008 to 4.2 ft/d in glacial sediments.

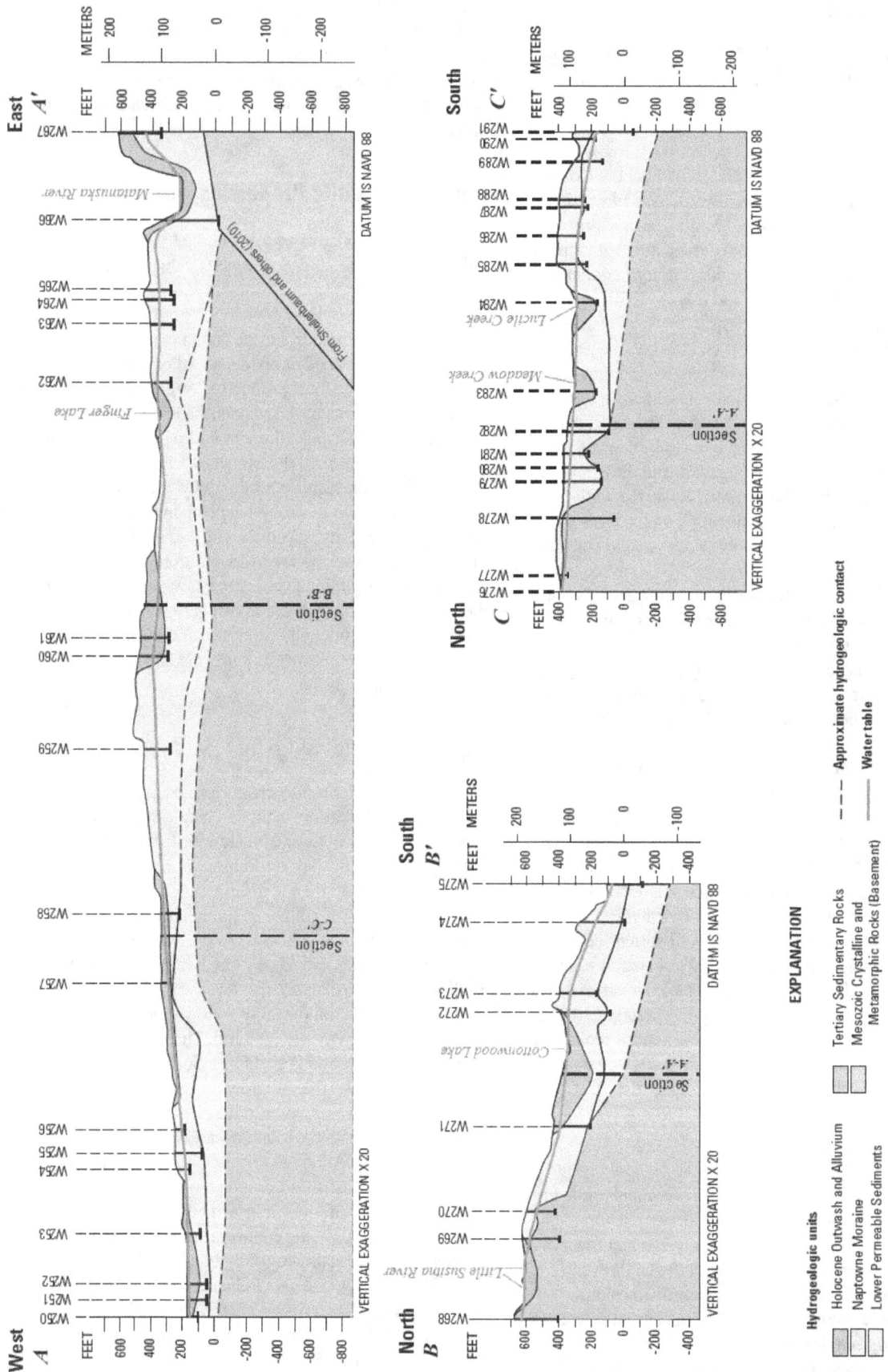

Figure 4. Matanuska-Susitna Valley, Alaska. (A) Section A-A'; (B) Section B-B'; (C) Section C-C'. Borehole data for wells are indexed in appendix A.

A. Base of Tertiary Sedimentary Rocks unit

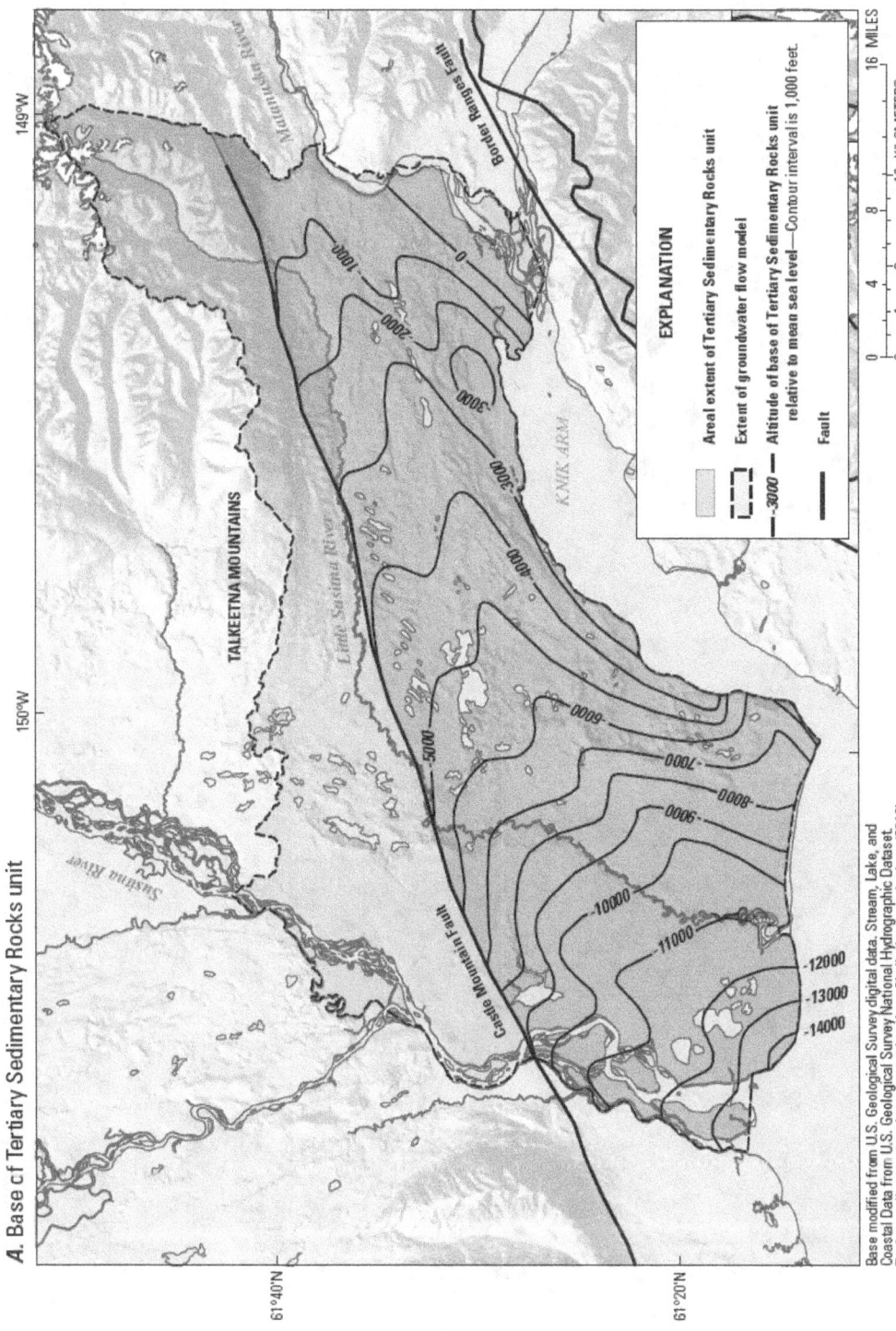

Base modified from U.S. Geological Survey digital data. Stream, Lake, and
Coastal Data from U.S. Geological Survey National Hydrographic Dataset.
Tertiary sedimentary rock contours taken from Shellenbaum and others (2010)

Figure 5. Extent and altitude of base of hydrogeologic units, Matanuska-Susitna Valley, Alaska. (*A*) Base of Tertiary Sedimentary Rocks;
(*B*) Base of Lower Permeable Sediments; (*C*) Base of Naptowne Moraine; (*D*) Base of Fine Sediments; (*E*) Base of Holocene Outwash and
Alluvium.

B. Base of Lower Permeable Sediments Unit

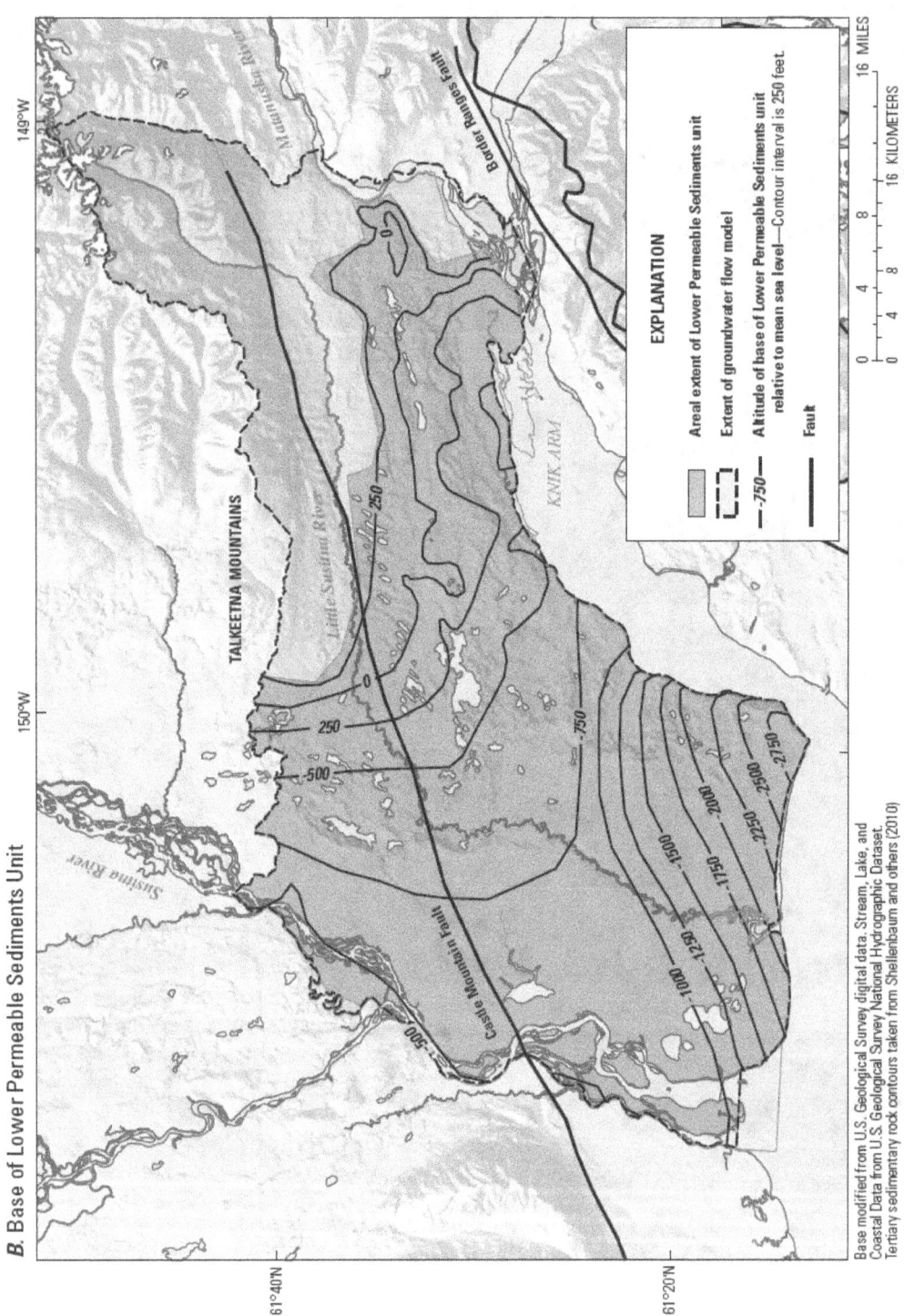

Base modified from U.S. Geological Survey digital data. Stream, Lake, and
Coastal Data from U.S. Geological Survey National Hydrographic Dataset.
Tertiary sedimentary rock contours taken from Shellenbaum and others (2010)

Figure 5.—Continued

C. Base of Naptowne Moraine unit

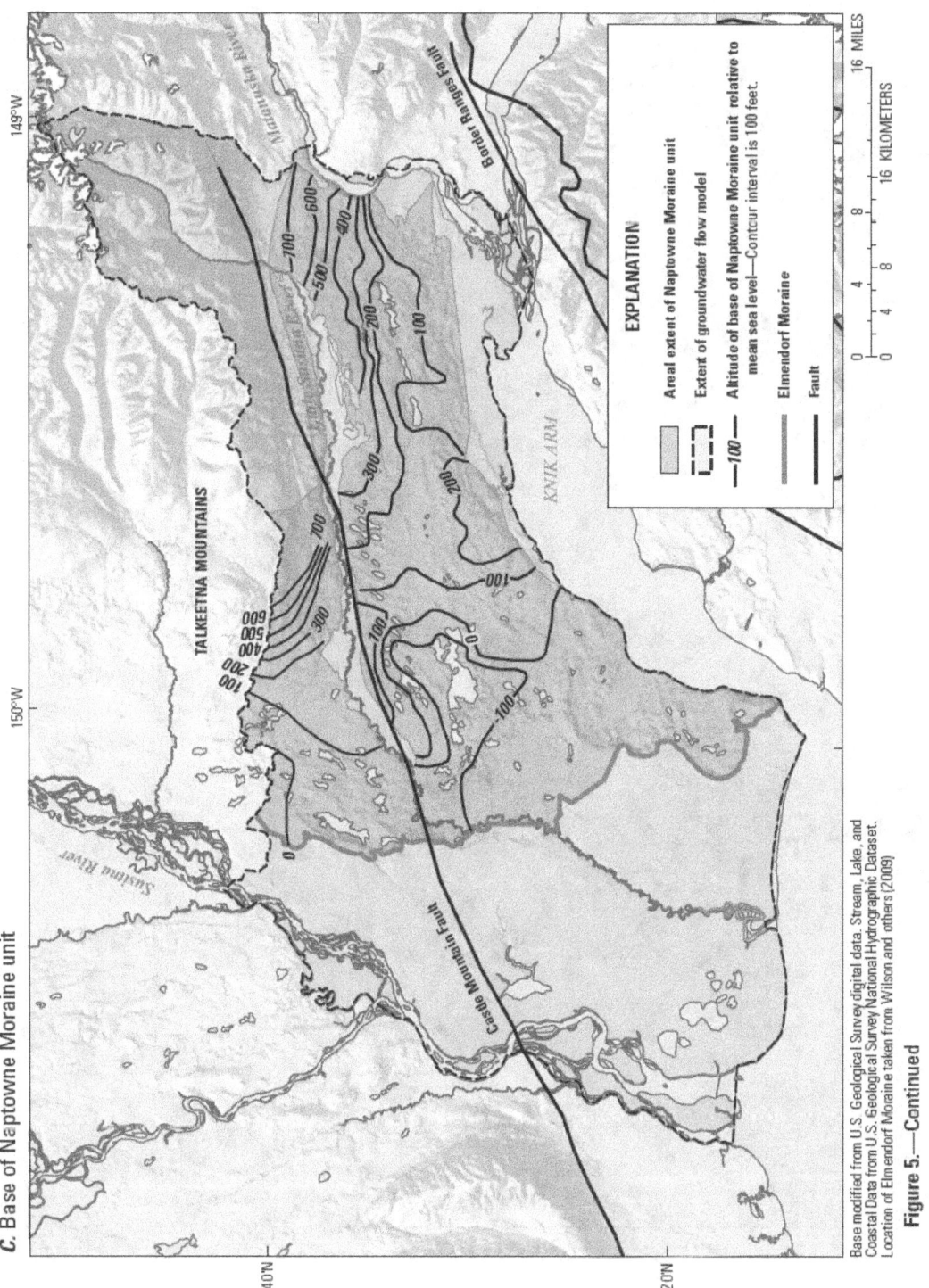

Base modified from U.S. Geological Survey digital data. Stream, Lake, and
Coastal Data from U.S. Geological Survey National Hydrographic Dataset.
Location of Elmendorf Moraine taken from Wilson and others (2009)

Figure 5.—Continued

D. Base of Fine Sediments unit

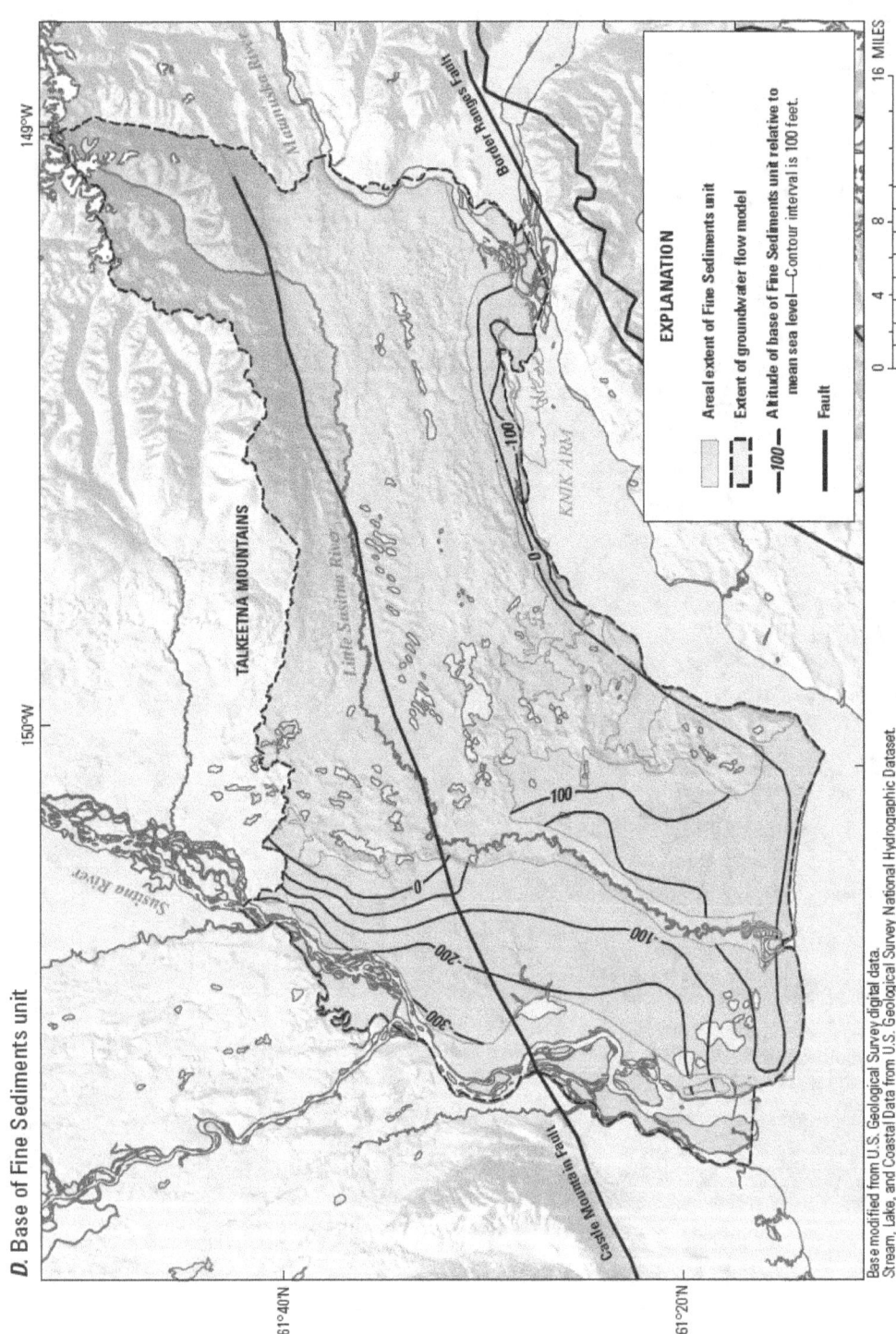

Base modified from U.S. Geological Survey digital data.
Stream, Lake, and Coastal Data from U.S. Geological Survey National Hydrographic Dataset.

Figure 5.—Continued

E. Base of Holocene Outwash and Alluvium unit

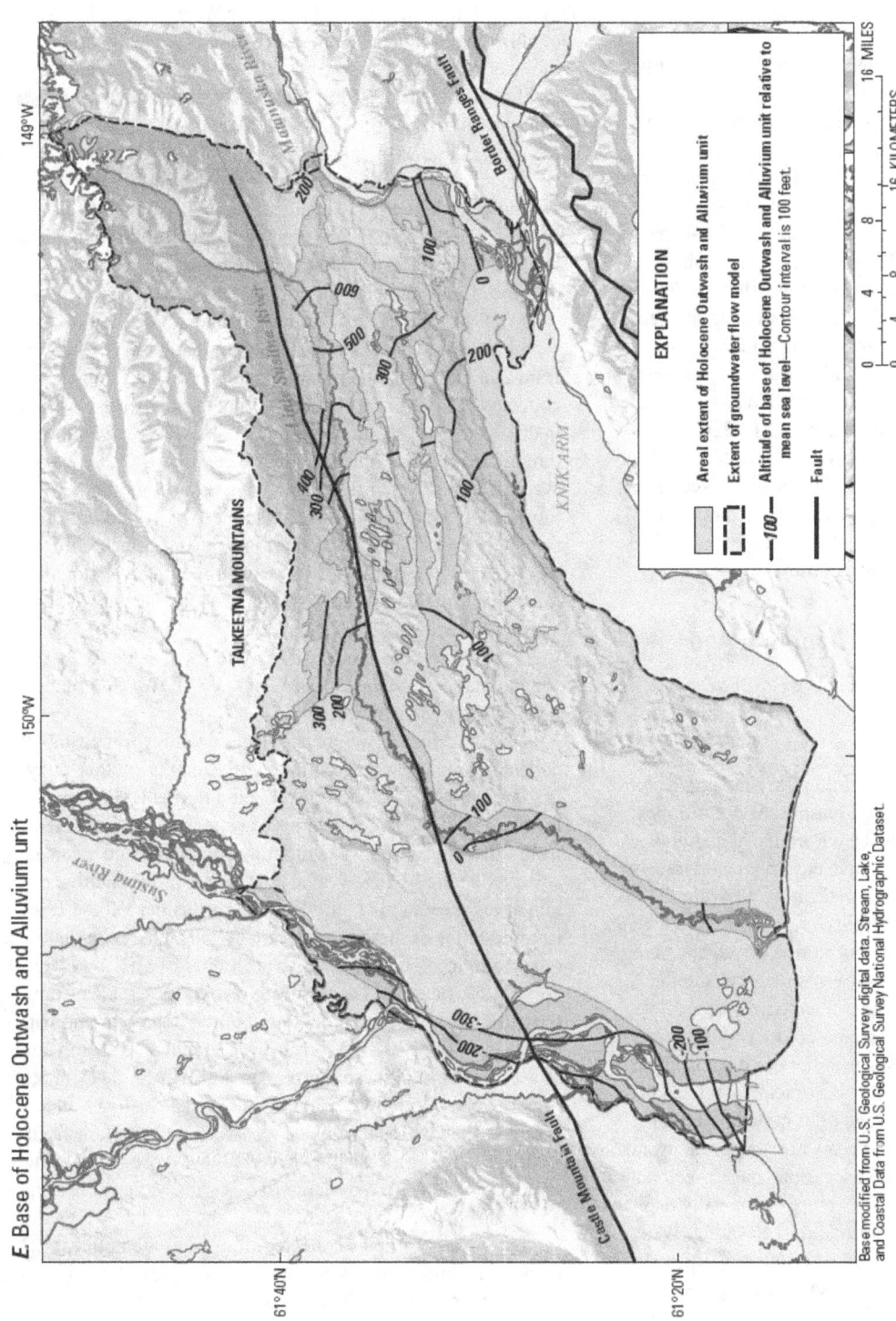

Base modified from U.S. Geological Survey digital data. Stream, Lake, and Coastal Data from U.S. Geological Survey National Hydrographic Dataset.

Figure 5.—Continued

Specific-Capacity Data

In the absence of aquifer-test data, specific-capacity data may be used to quantify well productivity, from which aquifer productivity may be inferred. The specific capacity, C_s, is defined as $C_s = Q/\Delta h_w$, where Q is the pumping rate and Δh_w is the drawdown in the well, typically recorded after 24 hours of continuous pumping (Driscoll, 1986). In the study area, well flow tests are required for municipal and high-capacity community wells, both of which are generally finished in very permeable unconsolidated sediments. Specific-capacity values calculated using well flow tests from municipal and community wells in the study area, along with relevant well-construction data and information from the flow test, are displayed in table 3. For wells finished in unconsolidated sediments, the median specific capacity was 1,226 ft³/d/ft. For aquifers with abundant data on specific capacity and transmissivity, empirical relations have been derived to relate those two properties (e.g., Razack and Huntley, 1991); however, transmissivity data are insufficient to undertake such analyses in the Matanuska-Susitna Valley.

Isotopic Composition and Apparent Age of Groundwater

Stable Isotopes of Water

Water isotopic ratios are useful in the characterization of hydrologic systems for several reasons. Water isotopes are chemically inert under most environmental conditions, making them useful as conservative environmental tracers. Also, water isotopic ratios vary predictably according to precipitation regimes. The isotopic composition of rain water changes predictably during rainout from a cloud, becoming more depleted in heavy isotopes over time. The isotopic depletion during rainout depends on temperature; greater isotopic depletion takes place during colder atmospheric conditions. Precipitation falling near the coast (an original moisture source) is generally less isotopically depleted than precipitation falling further inland or in mountain regions. The isotopic composition of surface water and groundwater can therefore be used to infer the location where precipitation fell because precipitation becomes more isotopically depleted as the cloud mass moves inland and up in elevation. Water isotopic ratios are typically evaluated by plotting the relation between the hydrogen isotope ratio and the oxygen isotope ratio. Globally, meteoric water isotopic compositions, on

average, fall along a Global Meteoric Water Line (GMWL) on this type of plot, as defined by Craig (1961). Deviations of meteoric water from the GMWL are observed locally, and Local Meteoric Water Lines (LMWL) are generally determined empirically and used in regional to local hydrologic studies.

Data on the isotopic composition of surface water and groundwater in the study area are available from several sources. Glass (2001) and Moran and Solin (2006) published analyses of surface-water and groundwater samples for deuterium:protium ($^2H:^1H$) and oxygen-18:oxygen-16 ($^{18}O:^{16}O$) isotopic ratios. In addition, event precipitation samples were collected at the Tideview station in Anchorage from 2007 to 2010, and monthly composite rain samples were collected at the MAES weather station during April-November in 2010 and 2011. Both precipitation datasets are available through the Alaska Water Isotopes Network (J. Welker, Alaska Water Isotopes Network [AKWIN], written commun., 2012). Surface-water and groundwater samples were collected during this study and analyzed for water isotopic ratios according to USGS protocols; the results were stored in the National Water Information System (NWIS). All isotopic data used in subsequent analyses are tabulated in appendix B. Hereafter, water isotopic ratios are reported as delta (δ) values relative to Vienna Standard Mean Ocean Water (VSMOW) (International Atomic Energy Agency, 2009).

The relation between water isotopic ratios from samples collected previously and samples collected during this study is shown in figure 6. The GMWL and inferred LMWL are included with the isotopic ratios of water samples collected from different sources. The monthly composite rain samples collected at the MAES weather station are representative of rain at low elevations in the Matanuska-Susitna Valley. The isotopic ratios of these samples are less depleted relative to other samples in this plot; the mean values of δ^2H and $\delta^{18}O$ are -111.5‰ and -14.1‰, respectively. Water samples from the Little Susitna River are representative of precipitation at higher elevations in the Talkeetna Mountains; the mean values of δ^2H and $\delta^{18}O$ in these samples are -142.5‰ and -18.6‰, respectively. Groundwater in the Matanuska-Susitna Valley is replenished by local in-place recharge and mountain-front recharge in the Little Susitna River Valley; most groundwater samples collected in the Matanuska-Susitna Valley constitute a mixture of these two sources (fig. 6). The δ^2H and $\delta^{18}O$ values for many water samples collected from lakes are noticeably higher than bulk monthly precipitation samples. Possible explanations for this difference are discussed below.

Table 3. Summary well construction data, details of well flow test, and calculated specific capacity values for selected wells , Matanuska-Susitna Valley, Alaska.

[All well data from the Alaska Department of Environmental Conservation Wasilla Office, Drinking Water Program, File Data. **Abbreviations:** NAD 83, North American Dataum of 1983; (ft³/d)/ft, cubic feet per day per foot; nd, no data; –, well location data not available]

Well No.	NAD 83 coordinates (latitude, longitude)	Depth of screened interval or depth of well opening (feet)	Aquifer material	Hydrogeologic unit	Discharge (gallons per minute)	Total drawdown (feet)	Total elapsed pumping time (hours)	Specific capacity [(ft³/d)/ft]
[1]W190	61°34'00"N., 149°41'10"W.	89–102	Gravel	Holocene Outwash and Alluvium	215	102	nd	4,869
W191	61°33'54"N., 149°17'42"W.	119.5–133	Gravel, gravelly sand	Lower Permeable Sediments	75	8.90	6.5	1,622
W192	61°32'05"N., 149°35'42"W.	141	Silty gravel, sand	Lower Permeable Sediments	5.5	2.60	4	407
		223–244	Sand and gravel		130	37.30	4	671
		218–228	Gravel, sand		290	20.00	4	2,791
W193	61°34'08"N., 149°20'51"W.	103	Sand and gravel	Holocene Outwash and Alluvium	5.5	4.00	4	265
W194	61°36'50"N., 149°20'51"W.	110–121	Gravel	Holocene Outwash and Alluvium	150	27.00	4	1,069
W195	61°36'06"N., 149°15'00"W.	68–88	Gravel	Holocene Outwash and Alluvium	15	61.92	7	47
W196	61°35'11"N., 149°20'34"W.	210–220	Gravel	Holocene Outwash and Alluvium	200	62.75	3.5	614
		71–76	nd		80	13.50	1	1,141
W197	61°35'25"N., 149°24'13"W.	115–125	Gravel	Holocene Outwash and Alluvium	65	57.50	6	218
		10	Sand		180	18.00	24	1,925
W198	61°39'17"N., 149°19'48"W.	40	Sandstone	Tertiary Sedimentary Rock	30	55.50	4	104
W199	61°36'47"N., 149°16'21"W.	48.5–174	Sandstone/ Bedrock	Tertiary Sedimentary Rock	31	116.00	12	51
		96–106	Gravel		150	13.00	4	2,221
W200	61°34'09"N., 149°09'28"W.	99.75–101.4	Gravel, sand and gravel	Holocene Outwash and Alluvium	170	3.25	9	10,069
W202	61°34'27"N., 149°22'30"W.	235–240	Sand and gravel	Naptowne Moraine	27	12.00	8.5	433
W203	61°36'30"N., 149°21'34"W.	90–101	Sand and gravel	Holocene Outwash and Alluvium	190	1.56	24	23,446
W204	61°35'06"N., 149°19'22"W.	68–74, 100–115	Sand and gravel	Holocene Outwash and Alluvium	15	11.00	4	263
W205	61°34'45"N., 149°23'53"W.	55 –58	Gravel	Holocene Outwash and Alluvium	6.3	9.00	4	135
W206	61°34'56"N., 149°37'54"W.	91–101	Gravel	Holocene Outwash and Alluvium	220	9.00	4	4,706
W207	61°36'05"N., 149°12'56"W.	64 –74	Sand and gravel	Naptowne Moraine	113	25.30	12	860

Table 3. Summary well construction data, details of well flow test, and calculated specific capacity values for selected wells , Matanuska-Susitna Valley, Alaska.—Continued

[All well data from the Alaska Department of Environmental Conservation Wasilla Office, Drinking Water Program, File Data. **Abbreviations:** NAD 83, North American Dataum of 1983; (ft^3/d)/ft, cubic feet per day per foot; nd, no data; –, well location data not available]

Well No.	NAD 83 coordinates (latitude, longitude)	Depth of screened interval or depth of well opening (feet)	Aquifer material	Hydrogeologic unit	Discharge (gallons per minute)	Total drawdown (feet)	Total elapsed pumping time (hours)	Specific capacity [(ft^3/d)/ft]
W208	61°34'42"N., 149°16'42"W.	64–74	Gravel and sand	Holocene Outwash and Alluvium	220	2.98	24	14,211
W209	61°35'19"N., 149°24'15"W.	116	Sandy gravel	Holocene Outwash and Alluvium	50	1.30	4	7,404
–	–	90–100	Gravel and sand	–	145	8.00	4	3,489
–	–	140	Sand and gravel	–	18	74.00	4	47
–	–	131	Gravel	–	35	15.00	4	449
–	–	91–101	nd	–	66	79.00	24	161
–	–	191–201	Gravel	–	42	13.00	12	622
W211	61°31'52"N., 149°50'22"W.	99–109	Gravel	Naptowne Moraine	200	43.50	4	885
W212	61°35'57"N., 149°21'21"W.	nd	Gravel	Holocene Outwash and Zlluvium	15	4.00	4	722
W213	61°36'02"N., 149°14'32"W.	37–42	Silty gravel	Holocene Outwash and Alluvium	30	12.00	4	481
–	–	97–119	Sand	–	150	2.44	4.25	11,834
W214	61°38'32"N., 149°19'01"W.	83–89	Sand and gravel	Naptowne Moraine	50	5.00	4	1,925
–	–	156–166	Gravel	–	160	6.00	4	5,133
–	–	nd	nd	–	87	7.00	4	2,393
W215	61°35'26"N., 149°30'02"W.	nd	nd	Naptowne Moraine	300	47.12	12	1,226
W216	61°36'27"N., 149°24'31"W.	198	Sand silt and gravel	Naptowne Moraine	10	105.00	4	18
–	–	353	Gravel	–	64	232.00	9	53
W217	61°37'33"N., 149°27'21"W.	46–56	nd	Tertiary Sedimentary Rocks	150	15.00	4	1,925
W218	61°34'36"N., 149°09'00"W.	110–121	Gravel	Holocene Outwash and Alluvium	155	0.05	4	596,750
W219	61°34'48"N., 149°14'29"W.	236–246	nd	Lower Permeable Sediments	181	14.00	4	2,489
W220	61°32'58"N., 149°47'39"W.	nd	nd	Holocene Outwash and Alluvium	5.3	0.30	4	3,401

[1]Alaska Department of Environmental Conservation Wasilla Office, Drinking Water Program, File Data.

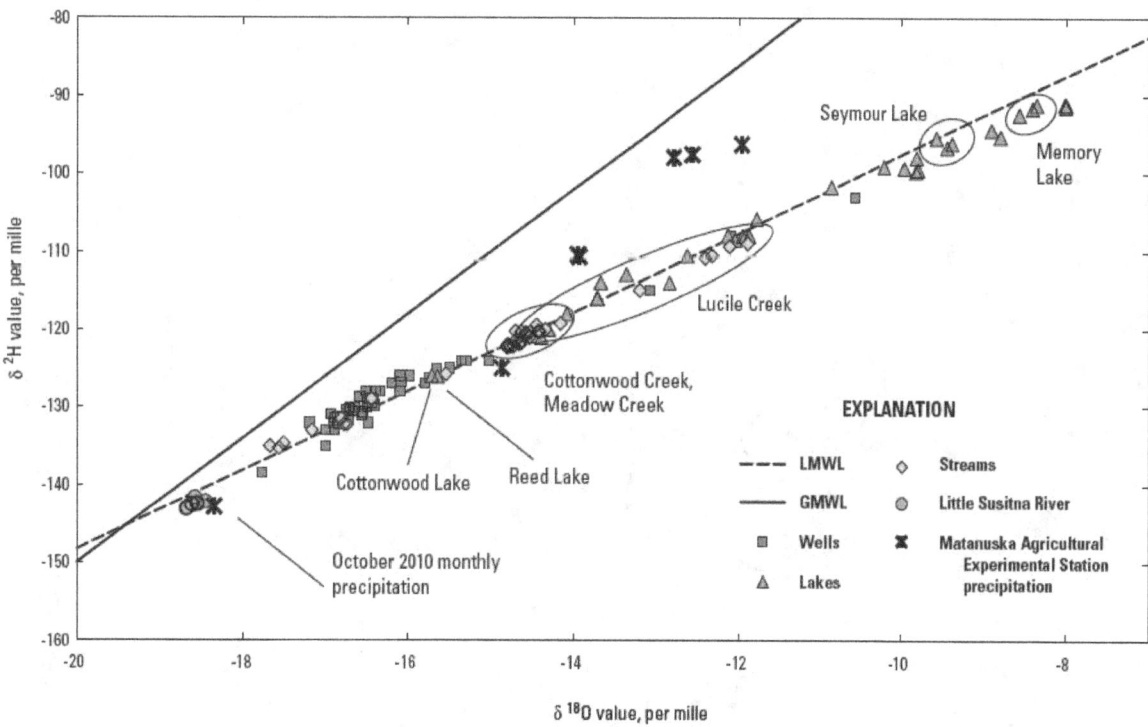

Figure 6. Isotopic composition of precipitation, surface-water, and groundwater samples collected in the Matanuska-Susitna Valley, Alaska, 1999–2011. Global meteoric water line (GMWL) and local meteoric water line (LMWL) are shown for reference.

Meteoric waters are expected to become increasingly depleted in heavier isotopes accompanying rainout in cloud masses moving north from Knik Arm into the Talkeetna Mountains. However, variability in the isotopic composition of surface water and groundwater does not follow this expected spatial pattern. Samples of particular water types (e.g., groundwater, lakes, etc.) were isotopically more similar than samples collected at similar elevations (fig. 7). The $\delta^{18}O$ values fell within the range of -15.0‰ to -17.0‰ for most of the groundwater samples. Three wells sampled near the Matanuska River had much lower $\delta^{18}O$ values of -20.0‰ to -21.3.0‰. The Matanuska River is glacially fed, originating as meltwater from the Matanuska Glacier. Isotopic analyses are not available for ice or meltwater from the Matanuska Glacier; however, water in ice cores from the nearby Eklutna Glacier had a mean $\delta^{18}O$ value of -20.1‰ (University of Alaska Anchorage, 2012). Water samples depleted in heavy isotopes are consistent with colder climate during recent glaciations. Surface-water samples collected from a side channel of the Matanuska River were similar in isotopic composition to these groundwater samples. It is likely these groundwater samples are representative of an alluvial aquifer in hydraulic connection with the Matanuska River.

In general, water samples from lakes were enriched in heavier isotopes compared to groundwater samples. This is particularly pronounced for lakes in the Meadow Lakes and Meadow Creek Area. The isotopic composition of lake-water samples is similar to the isotopic composition of monthly bulk rain samples collected at the MAES weather station; however, more than half of the lakes sampled were enriched in heavy isotopes relative to rain samples. One explanation for this discrepancy is that the sources of water for these lakes were not correctly characterized. Because only seven monthly composite rain samples had been collected from the MAES weather station at the time of this report, it is possible the mean isotopic signature from these samples is not an accurate representation of average conditions; this would be the case if the isotopic composition of local precipitation varies from year to year or has changed historically. Also, snowmelt likely contributes a substantial amount of in-place recharge to lakes, but data on the isotopic composition of snow in the Matanuska-Susitna Valley were not available at the time of this study. Alternately, isotopic enrichment following evaporation may explain the isotopic composition of these lakes (Clark and Fritz, 1997). This effect could be important in the presence of high potential evaporation rates.

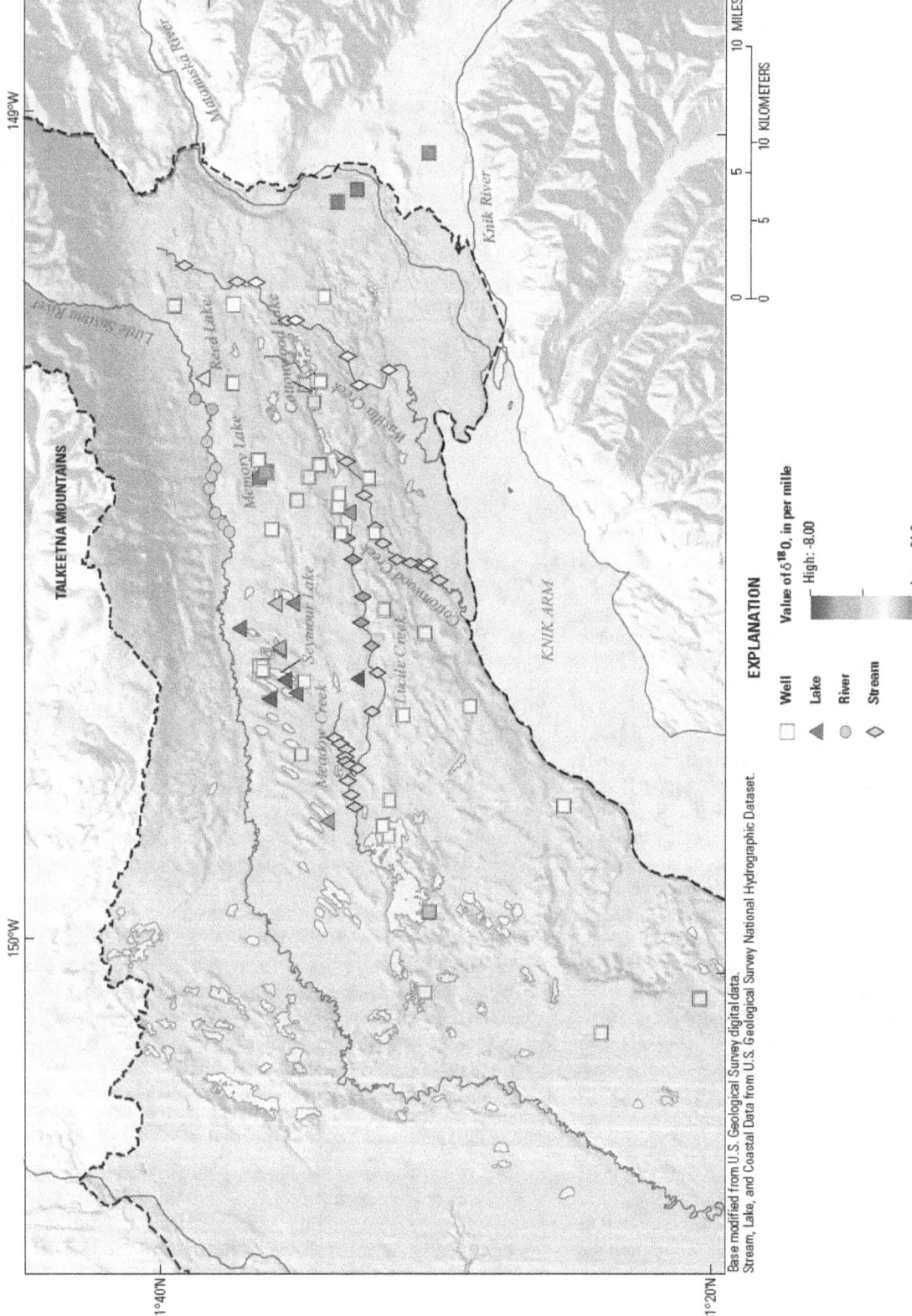

Base modified from U.S. Geological Survey digital data.
Stream, Lake, and Coastal Data from U.S. Geological Survey National Hydrographic Dataset.

Figure 7.   Distribution of $\delta^{18}O$ values from selected groundwater and surface-water sampling sites, Matanuska-Susitna Valley, Alaska.

Water samples from Cottonwood Lake and Reed Lake were depleted in heavy isotopes ($\delta^{18}O <$ -14.0‰) relative to other lake-water samples, indicating that these lakes may be hydraulically connected with the regional groundwater system. At Memory Lake, the groundwater from two wells downgradient from the lake is isotopically enriched relative to mean regional groundwater values, indicating leakage from the lake into the underlying aquifer. This is corroborated by similar major ion chemistry (Moran and Solin, 2006). In the area surrounding Seymour Lake, however, the isotopic composition of groundwater differed from the isotopic composition of nearby lakes. Moran and Solin (2006) proposed that isotopic fractionation during evaporation explains the different isotopic composition of water from lakes and nearby wells. This explanation is supported by similarities in major-ion geochemistry between Seymour Lake and nearby wells. Well driller's logs in the area surrounding Seymour Lake report layers of glacial till ranging in thickness from several feet to nearly 100 ft. Therefore, it is also possible that the groundwater samples collected near Seymour Lake are from aquifers hydraulically disconnected from the lake.

The locations of surface-water samples collected from Wasilla Creek, Cottonwood Creek, Lucile Creek, and Meadow Creek are shown in figure 7. Water samples collected from Meadow Creek and Cottonwood Creek were similar in isotopic composition; $\delta^{18}O$ values ranged from -14.8‰ to -14.3‰. The isotopic composition of water in Lucile Creek decreased steadily from -11.9‰ at the outlet of Lucile Lake to -14.8‰ at the confluence with Little Meadow Creek. Kikuchi and others (2012) attributed this longitudinal change in the isotopic composition of stream water to upwelling groundwater from a regional aquifer. Water samples collected along Wasilla Creek are depleted in heavy isotopes of water relative to the three other streams and isotopically resemble groundwater. This is not surprising because the headwaters of Wasilla Creek are very close to the Little Susitna River Valley, where a large amount of mountain-front recharge is thought to take place.

Groundwater Ages

Tritium and CFCs are commonly used as tracers for determining the apparent age of young groundwater. Glass (2001) presented water-quality data from 34 water wells in Upper Cook Inlet, 16 of which are in the Matanuska-Susitna Valley; groundwater samples collected from each well were analyzed for a variety of constituents, including tritium and CFCs. Interpretation of these data indicated that the groundwater residence time ranged from less than 1 to almost 60 years in the samples collected. For the most part, the spatial distribution of groundwater apparent ages (fig. 8) is consistent with the conceptual understanding of the groundwater system. Groundwater samples with the most recent probable recharge dates are typically located in proximity to surface-water features, indicating active exchange between groundwater and surface water. Three of the five samples with apparent ages

less than 7 years were collected from wells (W251, W255, and W256) finished in Holocene Outwash and Alluvium associated with the Matanuska River. For groundwater samples with apparent ages of at least 7 years, the apparent age increases in a west-southwest direction, from a recharge area in the Little Susitna River Valley towards the Susitna Lowlands. The two groundwater samples with apparent ages greater than 30 years are near Big Lake, a prominent surface-water feature; however, these samples were collected from deeper wells (W252 and W253) with open intervals beginning at least 100 ft below the land surface.

Aquifer Inflows and Outflows

Inflows

The Matanuska-Susitna Valley aquifer system is replenished by inflows from (1) in-place recharge, (2) inflows from surface-water bodies, (3) return flows from septic systems, and (4) irrigation return flows. In-place recharge was estimated on a monthly time step using a lumped-parameter land-surface model, spatially aggregated over watersheds within the study area. Surface-water infiltration to the aquifer system was investigated using seepage studies during base-flow conditions. Groundwater inflows from domestic septic systems and municipal drain fields were estimated through analysis of geospatial data, including parcel distribution and the extent of sewer lines. Irrigation return flows were estimated on the basis of landscape irrigation efficiency coefficients for irrigated properties.

In-Place Recharge

Precipitation from rainfall or snowmelt that does not evaporate or directly enter bodies of surface water by overland flow percolates through the unsaturated zone. Of this water, some is consumed by plants and some flows laterally toward bodies of surface water. The remainder continues percolating downward into the saturated zone and is referred to as in-place recharge. These processes are commonly treated in a one-dimensional framework but are applied over spatial scales ranging from plots to basins. A one-dimensional land-surface model called the Deep Percolation Model (DPM; Vaccaro and Maloy, 2006) was used to calculate in-place recharge over the study area. The DPM computes a water balance for hydrologic response units (HRUs) of arbitrary surface area on a daily time step, using time-series inputs of daily precipitation and minimum and maximum temperature. Precipitation inputs are partitioned at the land surface between shallow infiltration and evapotranspiration on the basis of user-supplied information about location, topography, land use, land cover, and soil hydraulic properties. The modeled water budget is then evaluated by comparing computed direct runoff and interflow with observed streamflow for the hydrologic unit. This model

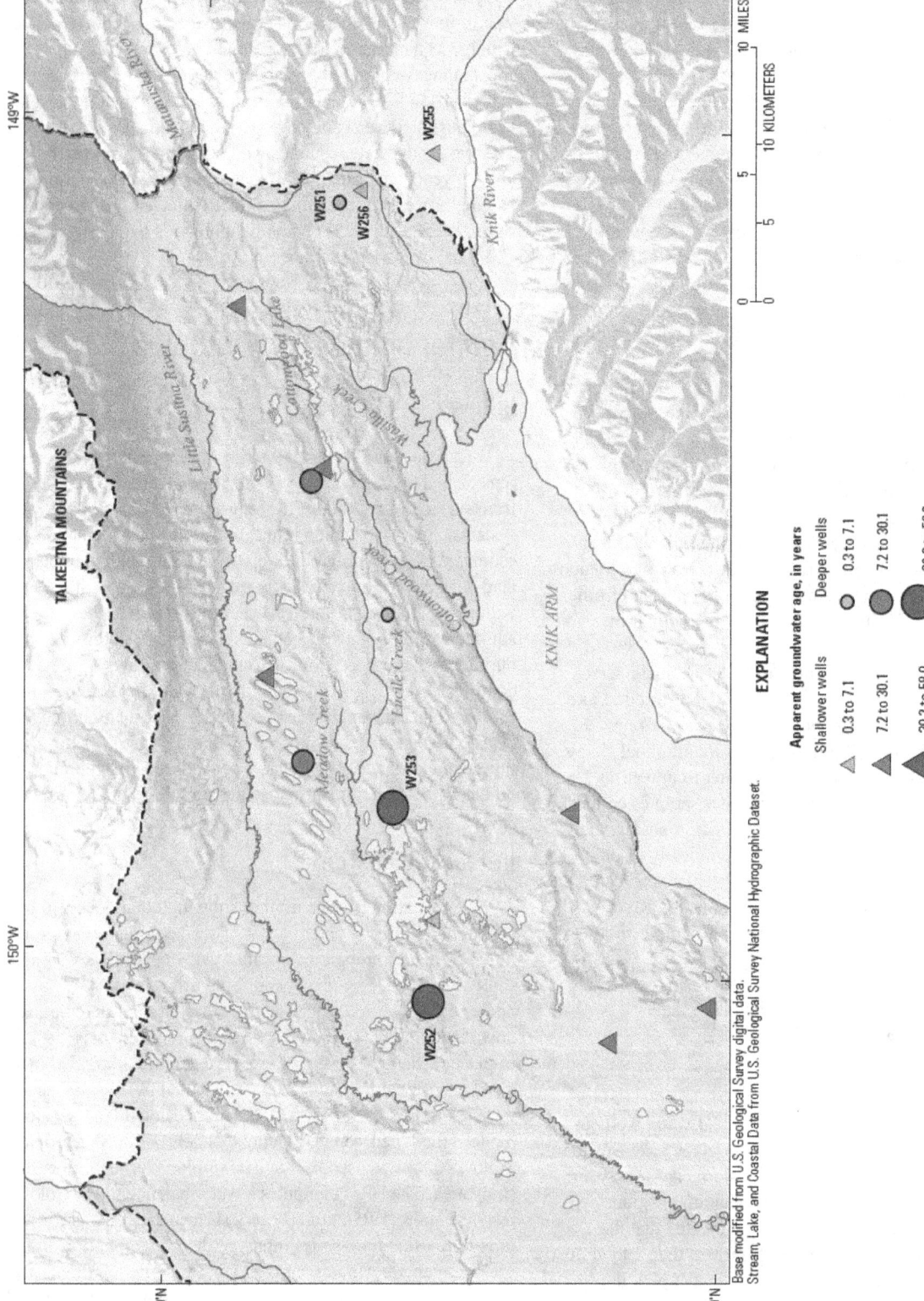

Figure 8.   Distribution of apparent groundwater ages estimated using tritium and helium-3 concentrations, Matanuska-Susitna Valley, Alaska. [Data from Glass (2001).]

was designed principally for application at the regional scale and was well-suited for calculating in-place recharge in the Matanuska-Susitna Valley.

HUC-12 hydrologic units from the Watershed Boundary Dataset (U.S. Department of Agriculture, Natural Resources Conservation Service, 2011c) were selected as the HRUs for this analysis of in-place recharge. There are 27 HUC-12 units within the boundaries of the study area (fig. 9). Some of these units are self-contained watersheds with one well-defined drainage; others are subdivisions of a larger watershed. For example, the Wasilla Creek HUC-12 (190204021302)has only one drainage, Wasilla Creek; on the other hand, the Little Susitna River watershed is divided into seven HUC-12 units. Selecting HUC-12s as the HRUs was advantageous because time-series streamflow data are available from USGS streamgages for some of the rivers and streams draining each unit, enabling comparison of direct runoff calculated by the model with observed values. The DPM requires basic physical information for each HRU, including location, slope, aspect, surface area, land use, soil type, and soil properties. Topographic data for each HRU were calculated using the Alaska 60 m DEM (U.S. Geological Survey, 2011a), land-use and land-cover data were extracted from raster datasets of the National Seamless Map (U.S. Geological Survey, 2011b), and the spatial extent of different soil types was obtained from the U.S. General Soils Map, STATSGO2 (U.S. Department of Agriculture, Natural Resources Conservation Service,

2011d). The required data for each HRU were compiled in a Geographic Information System (GIS), and the results are tabulated in appendix C.

Weather data are available from a network of National Climatic Data Center Cooperative Observer sites across the Matanuska-Susitna Valley (fig. 9). The reported period of record for many of the stations in the study area spans tens of years. However, for most of those stations, daily surface data are sparse, and segments of missing record range from days to years. The periods of record during the years 2000–2010 are sufficiently continuous—missing only days to 1 month of data —for six stations. The periods of record for daily surface data at those six stations—including precipitation and temperature —are summarized in table 4. Precipitation and temperature data from these sites were retrieved from the National Climatic Data Center (National Oceanic and Atmospheric Administration, 2011). Meteorological and snow data were retrieved for the Independence Mine SNOTEL site (U.S. Department of Agriculture, Natural Resources Conservation Service, 2011a) from 2007 to the present; these data were used as model inputs for HRUs contributing to the Little Susitna River. To generate the continuous time series required by DPM, it was necessary to estimate missing values in these datasets. To estimate missing temperature data, a set of linear regression equations was developed relating minimum and maximum temperatures between different weather stations (fig. 10A-B).

Table 4. Location, period of record, and statistics summarizing precipitation and temperature for weather stations used in the Deep Percolation Model, Matanuska-Susitna Valley, Alaska.

[**Abbreviations:** NCDC, National Climatic Data Center; NAD 83, North American Datum of 1983; –, not an NCDC cooperative station]

Station name	NCDC cooperative station identification No.	Latitude (NAD 83) (decimal degrees)	Longitude (NAD 83), decimal degrees	Elevation (feet above mean sea level)	Continuous period of record	Mean daily weather values, 2002–2010[1]		
						Mean annual precipitation (inches)	Mean value for minimum daily temperature (degrees Celsius)	Mean value for maximum daily temperature (degrees Celsius)
Matanuska Agricultural Experimental Station (MAES)	505733	61.57	-149.25	174	1984–2010	14.8	28.0	44.6
Palmer Municipal Airport (PMA)	506867	61.60	-149.10	243	2002–2010	10.2	29.1	44.9
Anderson Lake	500302	61.62	-149.33	449	1971–2010	19.4	28.9	44.5
Independence Mine SNOTEL (Site 1091)	–	61.78	-149.28	3,389	2006–2010	33.8	24.5	35.1
White's Crossing	509790	61.70	-150.00	251	1971–2009	18.0	23.5	43.3
Sutton	508915	61.72	-148.87	551	1978–2010	18.5	27.0	45.9

[1]All values are from data averaged over 2002–2010 with the exception of data from the Independence Mine SNOTEL site, which are averaged over 2007–2010.

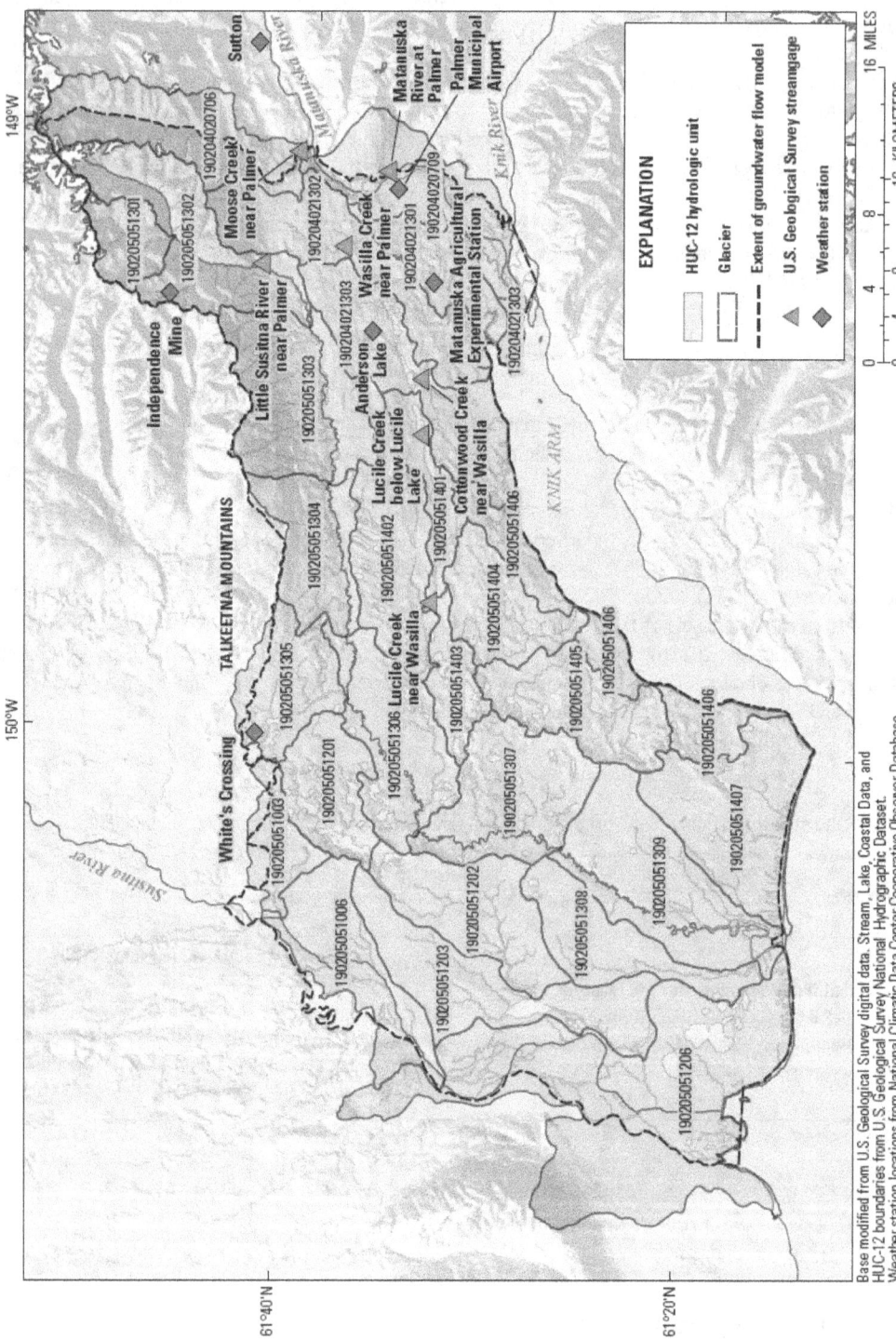

Base modified from U.S. Geological Survey digital data. Stream, Lake, Coastal Data, and
HUC-12 boundaries from U.S. Geological Survey National Hydrographic Dataset.
Weather station locations from National Climatic Data Center Cooperative Observer Database.

Figure 9.   HUC-12 boundaries, U.S. Geological Survey streamgages, and weather stations in the Matanuska-Susitna Valley, Alaska.

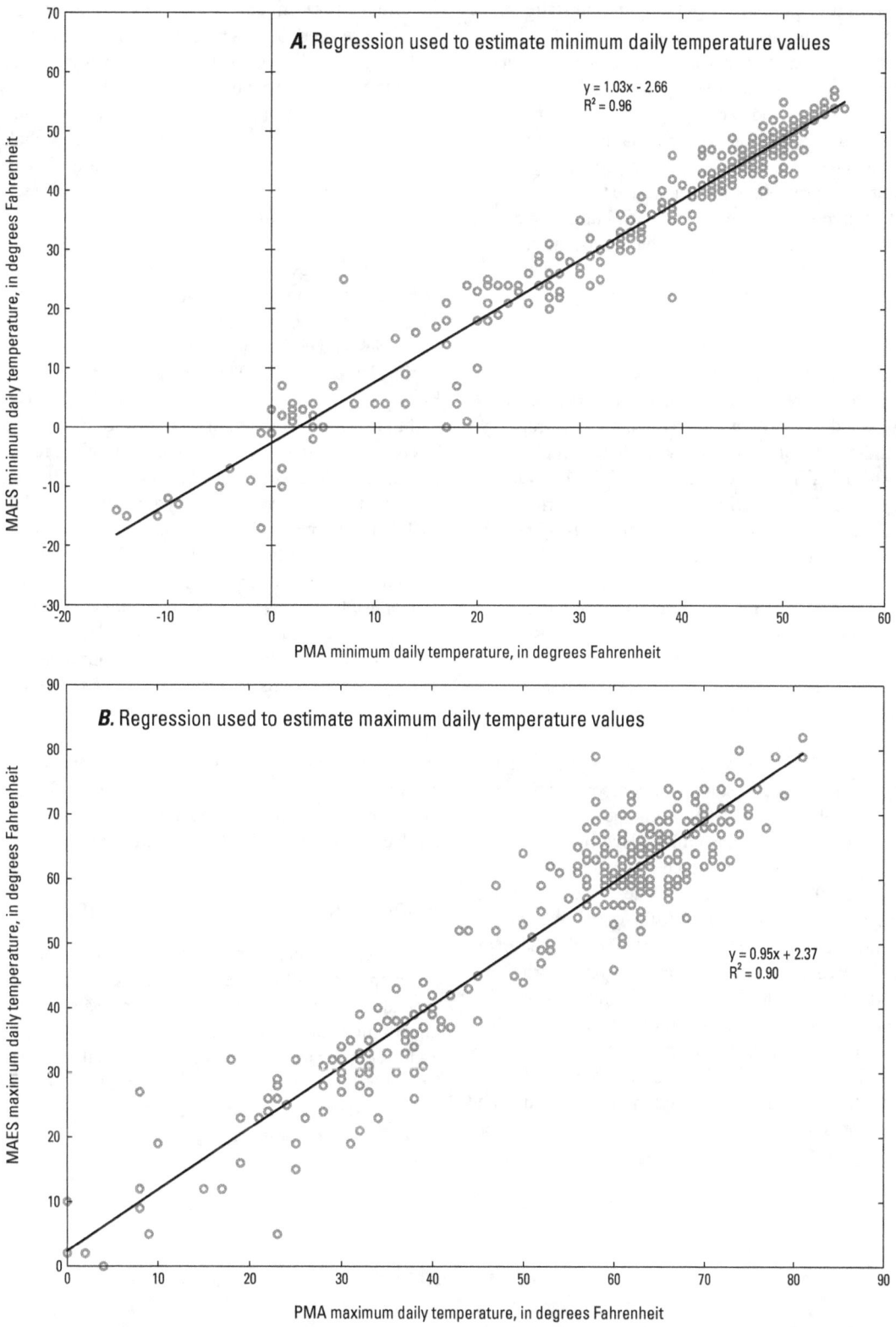

Figure 10. Linear regressions used to estimate missing temperature data at Matanuska Agricultural Experimental Station (MAES) weather station using data from Palmer Municipal Airport (PMA) weather station, Matanuska-Susitna Valley, Alaska. (*A*) Linear regression used to estimate minimum daily temperature values; (*B*) Linear regression used to estimate maximum daily temperature values.

The observed fit was quite strong ($R^2 > 0.9$ for all stations), inspiring confidence in this approach. A similar approach could not be used to estimate missing data in the precipitation time series because of spatial and temporal variability in precipitation across the study area. Therefore, missing precipitation data were estimated by calculating long-term mean daily values for the entire period of record. Daily values of temperature and precipitation are assigned to each HRU on the basis of an inverse distance-squared weighting procedure calculated internally within the DPM. The interpolation procedure also uses monthly basin average lapse rates to account for altitude-related differences between minimum and maximum daily temperatures.

In the study area, the soil becomes relatively impermeable during the ice-affected period. As a result, in-place groundwater recharge takes place primarily during the ice-free period. To account for the effects of soil freezing in the DPM, the value of the vertical saturated hydraulic conductivity (VKSAT) was set to vary with time. The value of VKSAT was set to zero during the ice-affected period (estimated as November through April). For HRUs without a streamgage at the catchment outlet, the value of the VKSAT during the ice-free period for each HRU was approximated using representative hydraulic properties of the dominant soil type (U.S. Department of Agriculture, Natural Resources Conservation Service, 2011b). For HRUs with a streamgage at the catchment outlet, the value of VKSAT was manually adjusted to minimize the difference between observed direct runoff and simulated direct runoff quantified in the DPM water budget. This approach also requires hydrograph separation into base-flow and stormflow components, so that the direct runoff calculated by DPM would be comparable to corresponding increases in streamflow. Hydrograph separation was accomplished using the computer program PART (Rutledge, 1998). This program is designed for use in hydrologic systems characterized by diffuse areal recharge that moves through shallow groundwater before discharging to a stream. The HRUs within the study area fit this description well, so the use of PART for hydrograph separation was deemed appropriate. Computation of direct runoff in DPM is performed under the assumption of a constant lag time from runoff generation to direct runoff arrival measured at USGS streamgages. On the basis of an inspection of precipitation and streamflow records for each HRU, this lag time value was set equal to zero.

Manual adjustment of the VKSAT parameter produced a close agreement between observed and simulated basin yield (fig. 11). The most suitable value was then used with DPM to calculate groundwater recharge on a monthly time step. Values for DPM parameters, along with detailed water budgets calculated by DPM for each HRU, are tabulated in appendix C. The annualized average recharge is highest in the mountainous HRUs—Moose Creek, Fishhook Creek-Little Susitna River, and Archangel Creek—in the northeastern corner of the study area (fig. 12). These HRUs receive annual precipitation in excess of 40 in., are characterized by thin, permeable soils and minimal vegetation, and are drained by Moose Creek and the Little Susitna River. Snowmelt in the late spring and early summer drives peak flows in the Little Susitna River (fig. 13). The simulated surface runoff calculated by DPM adequately matched these peak flows; more than half the annual precipitation remained for deep percolation. Therefore, it is likely that much of the in-place recharge leaves the Little Susitna River Valley as underflow. This finding is consistent with the conceptual hydrogeologic model presented by Jokela and others (1990), indicating that groundwater recharge in the Little Susitna River Valley drives regional groundwater flow patterns. In the core area, DPM calculated annual in-place recharge of approximately 5 in/yr. In-place recharge varied seasonally over the simulation period; most of the recharge takes place after snowmelt during the late spring or after seasonal rains characteristic of the late summer months (fig. 14). DPM results show that mean groundwater recharge rate over the entire study area is 6.1 in/yr. Multiplying the in-place recharge rate by the surface area for each HRU yields a volumetric in-place recharge rate for each HRU. Summing those values over the modeled area, the estimated volumetric in-place recharge rate is 260,000 acre-ft/yr.

Inflows from Surface-Water Bodies

Water flow between groundwater and surface water is driven by the differences in hydraulic head between groundwater and surface water. Surface water infiltrates into underlying aquifers when hydraulic head in the surface-water body is greater than the hydraulic head in the groundwater body. Flow from surface-water bodies into the groundwater system is discussed in the section, "Interaction Between Groundwater and Surface Water."

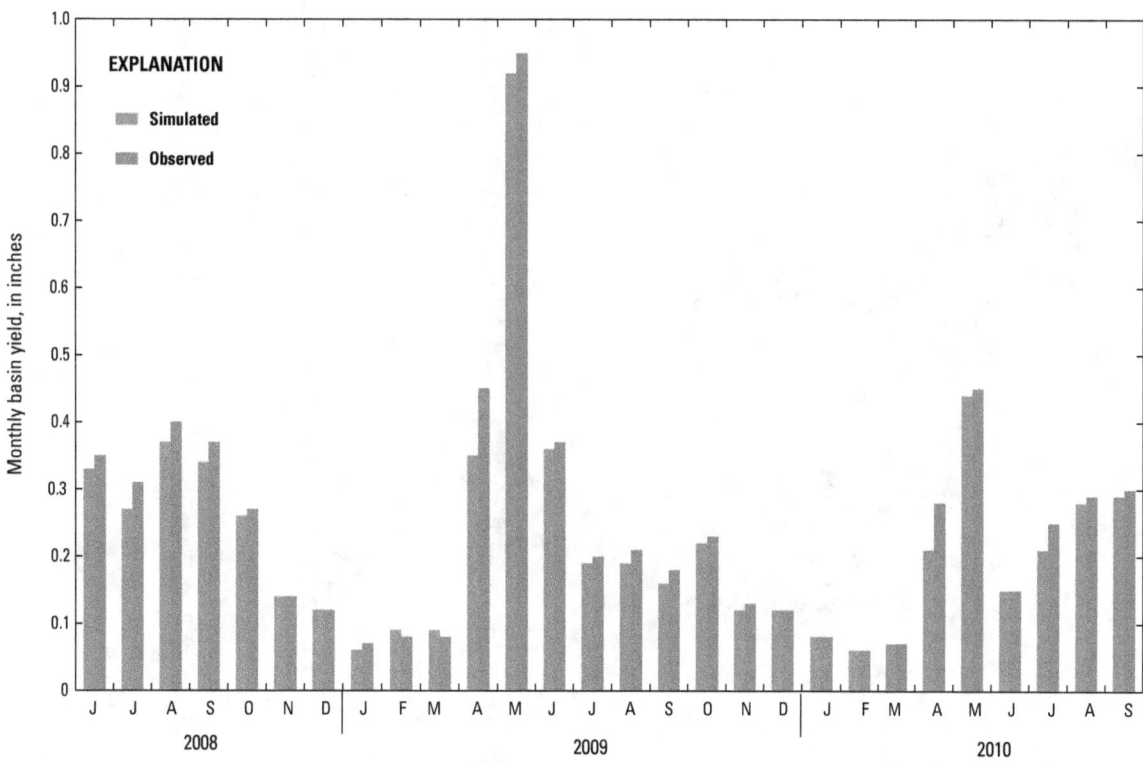

Figure 11.   Simulated and observed monthly basin yield for Wasilla Creek, Alaska, June 2008–September 2010.

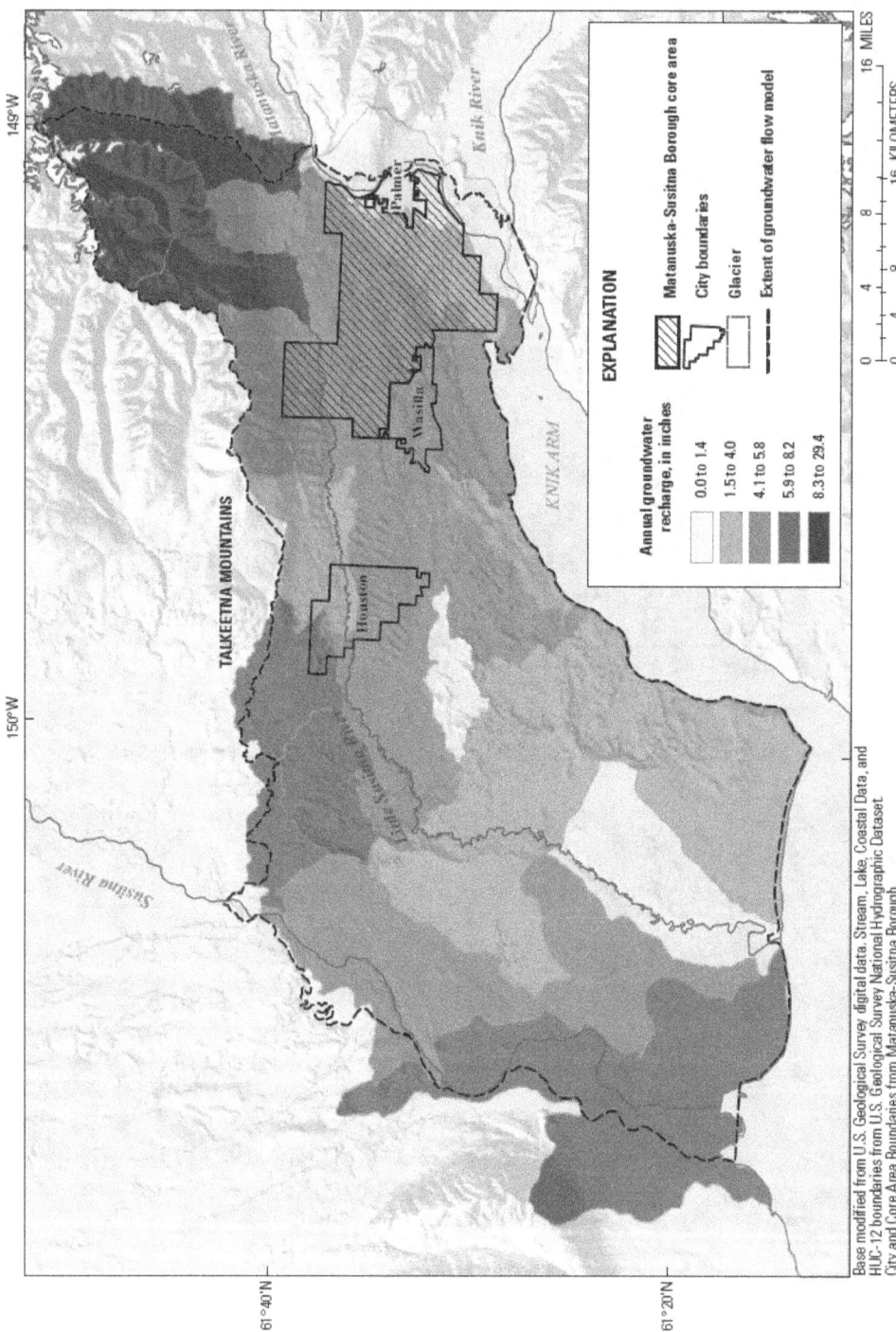

Base modified from U.S. Geological Survey digital data. Stream, Lake, Coastal Data, and
HUC-12 boundaries from U.S. Geological Survey National Hydrographic Dataset.
City and Core Area Boundaries from Matanuska-Susitna Borough.

Figure 12.   Distribution of annual groundwater recharge computed using the Deep Percolation Model (Vaccaro, 2007), Matanuska-Susitna
Valley, Alaska.

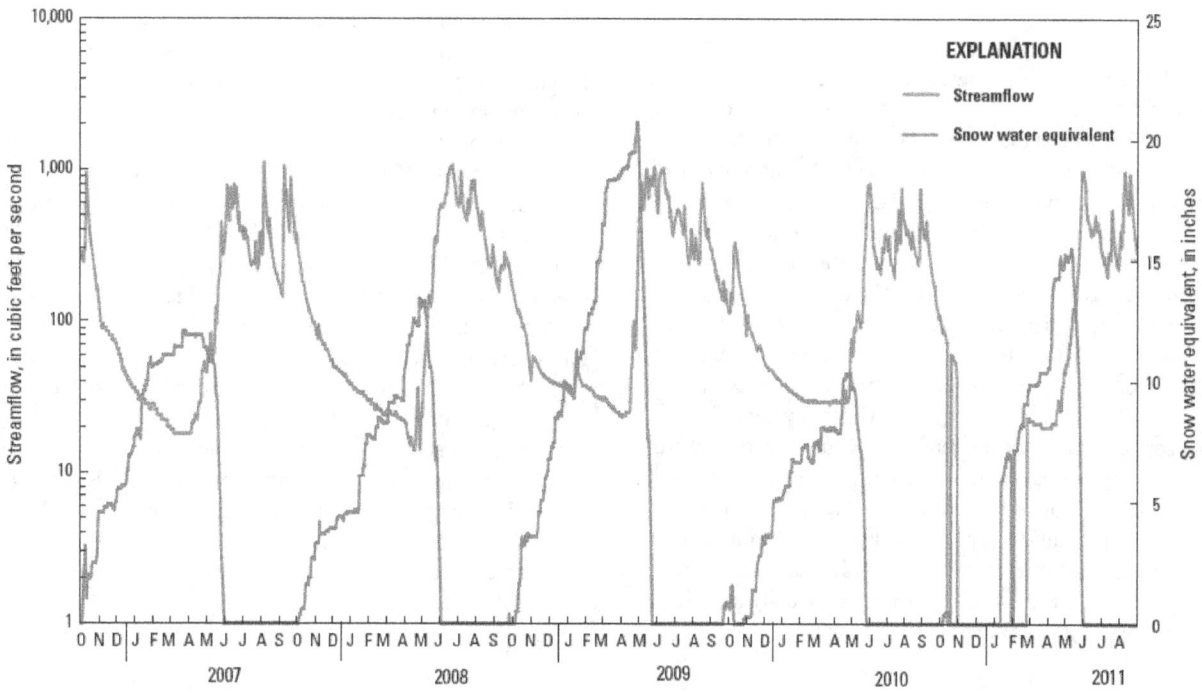

Figure 13. Snow water equivalent at the Independence Mine Snowpack Telemetry (SNOTEL) station and streamflow in the Little Susitna River, Alaska, October 2006–July 2011.

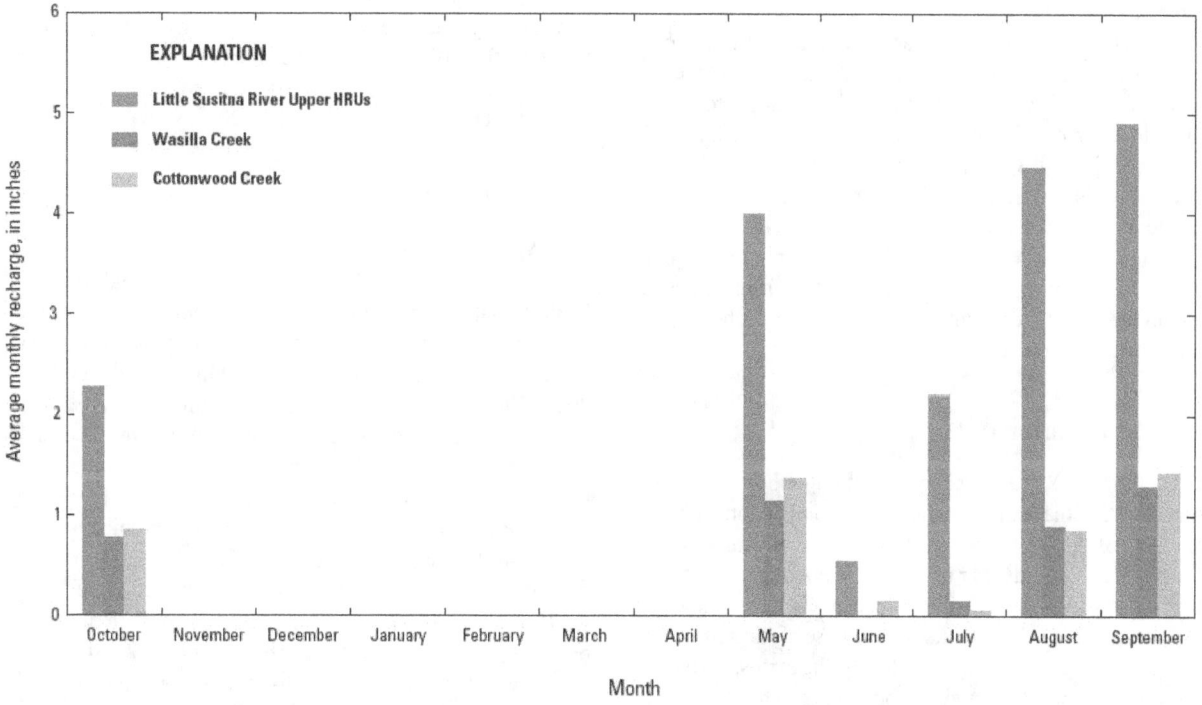

Figure 14. Average monthly in-place recharge, 2002–2010, for selected hydrologic response units (HRUs) in the Matanuska-Susitna Valley, Alaska. Little Susitna River Upper HRUs include Archangel Creek and Fishhook Creek-Little Susitna River.

Septic Effluent

Households outside the sewer service area for the cities of Wasilla and Palmer discharge wastewater to septic systems. A standard estimate of per capita water use in the Matanuska-Susitna Valley is 75 gal/d (J. MacInnis, Alaska Department of Environmental Conservation, oral commun., 2011). Ninety-five percent of that water use is assumed to become effluent discharged to septic systems that ultimately re-enters the groundwater system. The distribution of domestic septic systems was approximated by first identifying parcels in the study area with buildings. Those parcels within a 2,500-ft buffer of city sewer lines were assumed to receive sewer service and were excluded (fig. 15). Groundwater inflows from septic effluent for the remaining built-on parcels were calculated by multiplying the average per capita water use by 95 percent, multiplied by an estimated two persons per parcel; that is, 0.95 (75 gal/d × 2 persons). This calculation yields an estimated groundwater inflow rate of 142.5 gal/d per parcel. Using this groundwater inflow estimate, the daily areal inflow rate to a model grid cell was calculated as:

$$q_{septic} = \frac{1}{a_{parcels}}\left[n_{septic-parcels} \times 142.5 \frac{gallons}{parcel} \right] \quad (1)$$

In equation 1, $n_{septic-parcels}$ is the number of parcels in each model cell outside the sewer buffer area, and $a_{parcels}$ is the total area of those parcels.

In addition to effluent discharge from domestic septic systems, wastewater from households within the sewer service area for the city of Wasilla is treated and discharged to a drain field south of Wasilla Lake. From 2004 to 2010, the mean annual groundwater withdrawal by the city of Wasilla municipal wells was 850 acre-ft. Under the assumption that 95 percent of that water is discharged to the sewer system, the mean annual effluent discharge at the municipal drain field is 805 acre-ft. Summing the groundwater inflow from domestic and municipal septic effluent yields an estimated total volumetric inflow rate of 4,900 acre-ft/yr.

Irrigation Return Flows

On the whole, water use for irrigation in the Matanuska-Susitna Valley is minimal in comparison with other areas of the United States. Nonetheless, the majority of the metered groundwater use from water-supply wells in the Matanuska-Susitna Valley is in areas irrigated either for agriculture or for golf courses. For these areas, some percentage of the applied water is lost to the atmosphere by evapotranspiration; the remainder percolates beneath the plant root zone and eventually returns to the groundwater system. The percentage of applied water that is used by a cover crop is referred to as the irrigation efficiency. Data from the MAES weather station were used to estimate irrigation efficiency for the study area. The mean annual pan evaporation at the MAES weather station is 17.4 in. Assuming that potential evapotranspiration at this site is 70 percent of the pan evaporation yields an estimate of 12.2 in/yr for potential evapotranspiration. The irrigated acreage at the MAES site is approximately 260 acres, so the maximum loss from evapotranspiration at this site is 264 acre-ft/yr. The mean annual groundwater withdrawal at the MAES weather station from 2004 to 2010 was 570 acre-ft/year. Assuming that all this water was used for irrigation, the irrigation efficiency is 46 percent, meaning that 54 percent of the applied water ultimately returns to the groundwater system.

The irrigation efficiency of 46 percent was used to calculate groundwater recharge from irrigation return flows at irrigated parcels with metered groundwater withdrawals. Multiplying mean annual groundwater withdrawals at these parcels by the irrigation efficiency and summing over all irrigated parcels yield an estimated volumetric inflow rate of 1,900 acre-ft/yr.

Outflows

Outflows from the Matanuska-Susitna Valley aquifer system include (1) groundwater withdrawals from wells, (2) groundwater discharge to rivers, streams, and lakes, and (3) groundwater discharge to Knik Arm. Groundwater withdrawals from wells were estimated using pumpage records from high-capacity municipal and community wells in and surrounding Palmer and Wasilla. Groundwater discharge rates to the Little Susitna River and small streams were determined through field investigations of groundwater/surface-water interaction performed during 2010–2011.

Groundwater Withdrawals

All groundwater use in the Matanuska-Susitna Valley falls into one of three major categories: municipal or institutional supply, community well supply, and domestic users. Annual groundwater withdrawal records for municipal and community supply wells were obtained for the period 2004–2010 (Roy Ireland, Alaska Department of Natural Resources, written commun., October 2011). Information on the supply wells for which groundwater withdrawal records are available is shown in table 5; the locations of those supply wells are shown in figure 16. The depths of municipal and supply wells ranged from 48 to 663 ft, and most wells are finished in gravel or sandy gravel corresponding to either the Holocene Outwash and Alluvium or Lower Permeable Sediments hydrogeologic units. From 2004 to 2010, net annual groundwater use for individual supply wells ranged from less than 1 to more than 1,500 Mgal. The mean annual groundwater withdrawal rate for all municipal and community supply wells was 490 Mgal/yr, or 1,700 acre-ft/yr, during the 2004–2010 period.

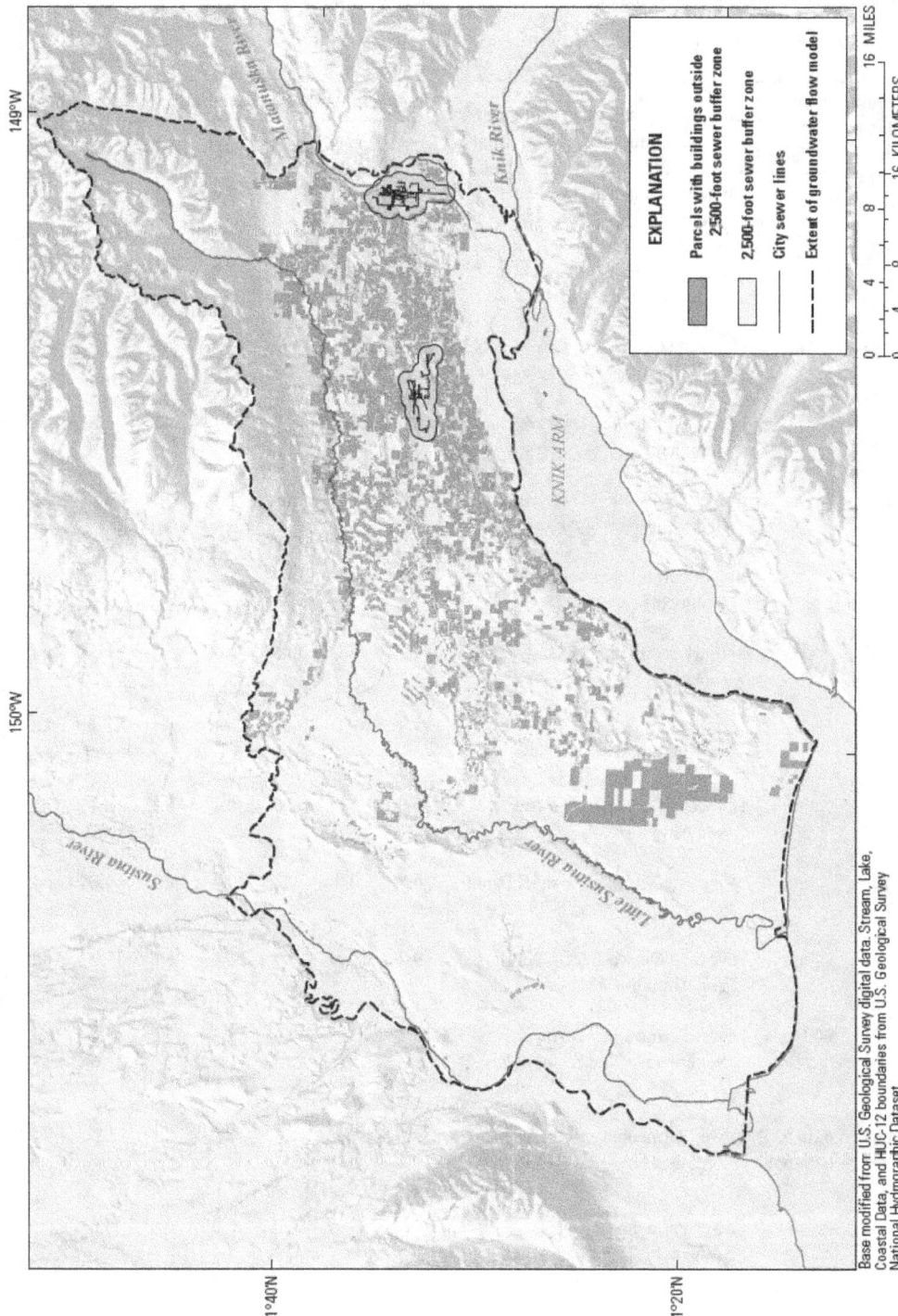

Base modified from U.S. Geological Survey digital data. Stream, Lake, Coastal Data, and HUC-12 boundaries from U.S. Geological Survey National Hydrographic Dataset.

Figure 15. Locations of parcels with buildings, city sewer lines for the cities of Palmer and Wasilla, and a 2,500-foot sewer buffer zone surrounding city sewer lines, Matanuska-Susitna Valley, Alaska.

Table 5.   Well construction and water use data for supply wells in the Matanuska-Susitna Valley, Alaska.

[**Abbreviations:** NAD 83, North American Dataum of 1983; na, not applicable; nd, no data; inc, incomplete record]

Well No.	NAD 83 coordinates (latitude, longitude)	Well depth (feet)	Screened interval (feet)	Hydrogeologic unit	Aquifer material	Annual groundwater withdrawal (million gallons)						
						2004	2005	2006	2007	2008	2009	2010
W221	61°34′02″N., 149°09′21″W.	100	na	Holocene Outwash and Alluvium	Gravel	5.0	1.7	3.5	3.2	2.6	nd	nd
W222	61°34′51″N., 149°15′42″W.	78	na	Holocene Outwash and Alluvium	Gravel	5.4	4.2	3.8	inc	nd	nd	nd
W223	61°34′51″N., 149°15′42″W.	119	na	Holocene Outwash and Alluvium	Gravel,cobbles	nd	nd	nd	nd	nd	nd	nd
W224	61°35′00″N., 149°17′01″W.	222	na	Lower Permeable Sediments	Gravel,sand	10.0	10.5	16.9	16.1	nd	inc	nd
W225	61°35′00″N., 149°17′01″W.	222	na	Lower Permeable Sediments	Gravel,sand	nd	nd	nd	nd	nd	nd	nd
W226	61°35′00″N., 149°17′01″W.	222	na	Lower Permeable Sediments	Gravel, sand	nd	nd	nd	nd	nd	nd	nd
W227	61°34′41″N., 149°16′51″W.	202	na	Tertiary sedimentary rocks	Sandstone	nd	nd	nd	nd	nd	nd	nd
W228	61°34′41″N., 149°16′51″W.	80	na	Holocene Outwash and Alluvium	Sand	nd	nd	nd	nd	nd	nd	nd
W229	61°36′24″N., 149°13′08″W.	115	na	Holocene Outwash and Alluvium	Gravel	0.0	0.0	0.0	0.1	0.1	0.1	nd
W230	61°36′05″N., 149°12′33″W.	100	na	Naptowne Moraine	Gravel (small)	0.8	0.0	0.2	0.1	0.0	0.0	nd
W231	61°36′18″N., 149°14′09″W.	48	na	Holocene Outwash and Alluvium	Gravel (large), sand	9.2	12.1	12.9	16.6	16.6	17.2	nd
W232	61°36′33″N., 149°08′50″W.	624	600–624	Lower Permeable Sediments	Gravel, boulders, sand	10.4	3.1	7.6	7.1	15.0	13.4	10.1
W233	61°36′29″N., 149°08′47″W.	663	620–635, 649–659	Lower Permeable Sediments	Gravel, sand (both coarse), gravel, sand	nd	nd	nd	nd	nd	nd	nd
W234	61°36′29″N., 149°08′44″W.	140	na	Lower Permeable Sediments	nd	0.0	0.0	0.0	0.0	0.0	0.0	0.0

Table 5. Well construction and water use data for supply wells in the Matanuska-Susitna Valley, Alaska.—Continued

[**Abbreviations:** nd, no data; inc, ncomplete record]

Well No.	NAD 83 coordinates (latitude, longitude)	Well depth (feet)	Screened interval (feet)	Hydrogeologic unit	Aquifer material	Annual groundwater withdrawal (million gallons)						
						2004	2005	2006	2007	2008	2009	2010
W235	61°34'19"N., 149°09'44"W.	100	90–100	Holocene Outwash and Alluvium	Gravel, sand	15.8	11.4	10.0	10.6	8.5	10.0	nd
W236	61°34'19"N., 149°09'44"W.	100	90–100	Holocene Outwash and Alluvium	Gravel, sand	nd	nd	nd	nd	nd	nd	nd
W237	61°30'21"N., 149°36'42"W.	159	144–159	Holocene Outwash and Alluvium	Gravel, sand, silt	nd	8.2	5.8	6.2	6.1	6.4	nd
W238	61°36'38"N., 149°13'19"W.	290	273–288	Lower Permeable Sediments	Silty, sandy gravel	7.7	2.3	292.5	298.1	277.3	240.5	nd
W239	61°33'58"N., 149°14'39"W.	138	118–138	Naptowne Moraine	Sand (coarse), gravel	11.6	nd	nd	13.3	12.5	0.0	nd
W240	61°31'24"N., 149°37'02"W.	152	152	Naptowne Moraine	Gravel, clay	20.9	25.3	19.5	88.8	82.9	82.6	52.5
W241	61°35'38"N., 149°29'38"W.	nd	na	nd	nd	28.1	37.0	29.0	0.4	3.1	10.7	34.1
W242	61°35'38"N., 149°29'38"W.	181	160–175	Naptowne Moraine	Gravel	1.9	2.0	1.8	2.2	1.9	1.5	1.4
W243	61°35'59"N., 149°26'04"W.	236	210–236	Naptowne Moraine	Gravel, sand	206.4	191.6	229.5	195.4	243.3	167.4	119.0
W244	61°35'57"N., 149°26'42"W.	281	252–263	Lower Permeable Sediments	Sand, gravel	nd	nd	4.8	1.8	2.2	0.2	0.0
W245	61°33'10"N., 149°19'34"W.	290	290–300	Lower Permeable Sediments	Sand, gravel	nd	nd	nd	4.0	4.5	0.3	0.4
W246	61°33'05"N., 149°18'51"W.	119	102–117	Lower Permeable Sediments	Sand, gravel	15.8	nd	11.8	14.0	6.6	12.9	nd

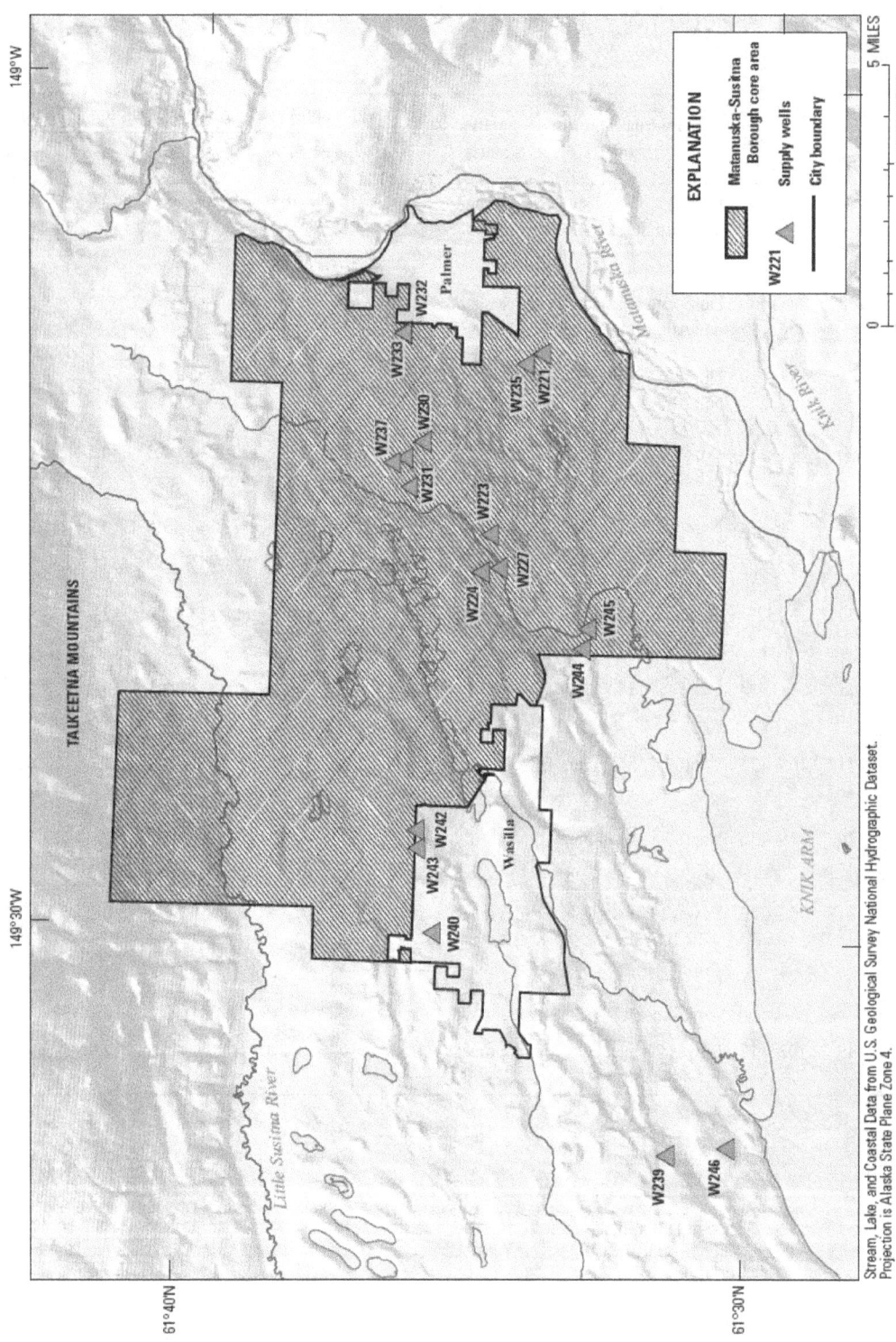

Stream, Lake, and Coastal Data from U.S. Geological Survey National Hydrographic Dataset.
Projection is Alaska State Plane Zone 4.

Figure 16.   Locations of water-supply wells with groundwater withdrawal data, Matanuska-Susitna Valley, Alaska, 2004–2010.

Data on groundwater withdrawals by domestic users outside municipal and community service areas are not currently available. Groundwater withdrawals by domestic users were estimated to be 150 gal/d per person for each parcel, assuming two people per parcel and an average per capita water use of 75 gal/d. The volumetric daily groundwater withdrawal rate for each model cell was then calculated as:

$$q_{domestic} = \left[n_{domestic-parcels} \times 150 \frac{gallons}{parcel} \right] \qquad (2)$$

The water distribution lines and sewer lines are similar in location and extent across the city of Wasilla service area. Therefore, it was assumed that parcels outside the sewer buffer area do not receive water from municipal wells. The value of $n_{domestic-parcels}$ in equation 2 is the number of parcels in each model cell outside the sewer buffer area and outside a subdivision supplied by a community well. Summing the estimated groundwater withdrawals by domestic wells over the entire model domain yielded a mean volumetric outflow rate of 4,100 acre-ft/yr. Combining groundwater withdrawals from domestic, community, and municipal wells, the mean annual volumetric outflow rate is 5,800 acre-ft/yr.

Outflows to Surface-Water Bodies

Groundwater discharges into surface-water bodies where the hydraulic head in the surface-water body is less than the hydraulic head in the groundwater body. If the head difference between surface water and groundwater and hydraulic conductivity of the aquifer material are large, flow rates between groundwater and surface water may be large. Outflow from groundwater into surface-water bodies is discussed in the section, "Interaction Between Groundwater and Surface Water."

Groundwater Discharge to Knik Arm

The water-table maps of Jokela and others (1990) and Moran and Solin (2006) show that at the regional scale, the water-table slopes south from the Talkeetna Mountains into Knik Arm and Cook Inlet. At the coast, groundwater is discharged to springs and seepage faces and as submarine groundwater directly into the ocean. Field data were insufficient to quantify groundwater outflow rates to Knik Arm; this is an important area for future study.

Interaction Between Groundwater and Surface Water

Interaction Between Groundwater and the Little Susitna River

The water-table map constructed by Jokela and others (1990) shows strong horizontal hydraulic gradients moving south from the Little Susitna River Valley into the core area, indicating that infiltration from the Little Susitna River may be an important source of inflow to the groundwater system. However, the water-table map alone does not conclusively determine whether these gradients are driven by underflow from the Little Susitna River Valley or from river seepage along the Little Susitna River. It was therefore deemed important to quantify groundwater inflow from the Little Susitna River.

To quantify groundwater inflow from the Little Susitna River, synoptic differential streamflow measurements (seepage runs) were conducted along the river during base-flow conditions in September 2010 and October 2011. In a seepage run, nearly instantaneous streamflow measurements are made along a stream. A systematic increase in streamflow for downstream measurements indicates groundwater seepage to the stream, and a systematic decrease in streamflow for downstream measurements indicates infiltration of stream water to the groundwater system. The locations of streamflow measurements made during seepage runs and USGS streamgages are shown in figure 17. Streamflow during the October 2011 seepage run was approximately 100–150 ft³/s less than during the September 2010 seepage run. The data from both seepage runs show a gradual increase in streamflow along the river (table 6). A slight decrease in streamflow was observed between the USGS streamgage 15290000 and Edgerton Parks Road during the October 2011 seepage run; however, this difference fell within the measurement error. The Little Susitna River received surface-water inflows from numerous tributary streams draining the Talkeetna Mountains along the study reach for the 2010–2011 seepage runs. It was not possible to measure streamflow in these tributary streams. Therefore, it is not possible to distinguish between surface-water and groundwater inflows to the river. It is possible that streamflow losses over the study reach of the Little Susitna River were masked by surface-water inflow from the tributary streams; however, existing data are insufficient to address this possibility.

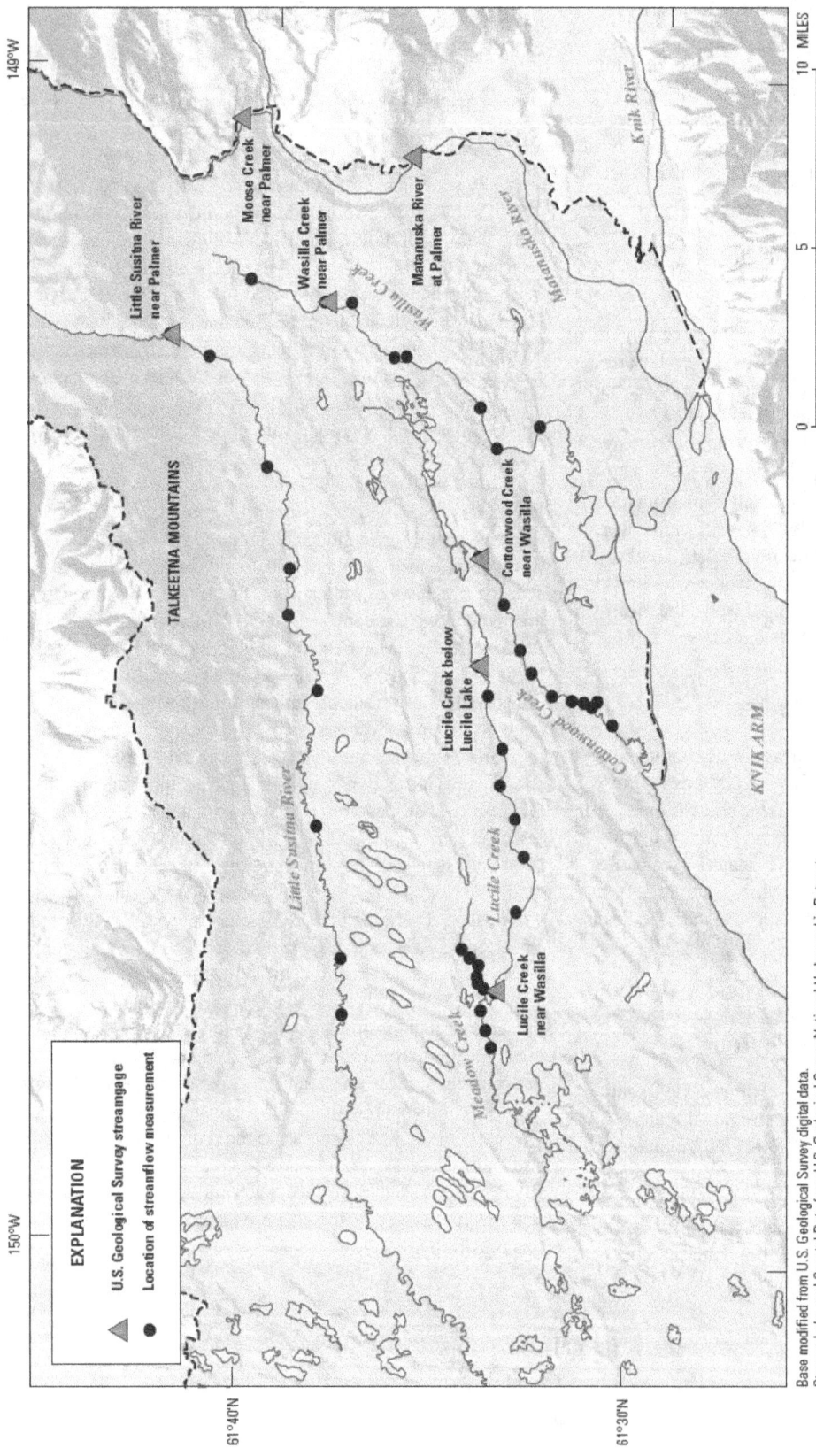

Base modified from U.S. Geological Survey digital data.
Stream, Lake, and Coastal Data from U.S. Geological Survey National Hydrographic Dataset.

Figure 17. Locations of U.S. Geological Survey streamgages and instantaneous streamflow measurements made during seepage runs, Matanuska-Susitna Valley, Alaska, 2009–2011.

Table 6. Streamflow measurements, specific conductance, and chloride concentrations for sites along the Little Susitna River, Alaska.

[Streamflow measurements are used to determine streamflow gains (positive values) and streamflow losses (negative values) between measurement sites. **Abbreviations:** ft³/s, cubic feet per second; μS/cm, microsiemens per centimeter; mg/L, milligram per liter; nd, no data; –, not calculated]

Measurement site or streamgage	Map identification No.	River mile	September 21, 2010		October 20, 2011		October 10–11, 2011	
			Streamflow (ft³/s)	Gain or loss (ft³/s)	Streamflow (ft³/s)	Gain or loss (ft³/s)	Specific conductance (μS/cm)	Chloride (mg/L)
Little Susitna River near Palmer (15290000)	LS-01	0.0	162.0	–	52	–	106	7.49
Little Susitna River at Edgerton Parks Road	LS-02	1.4	nd	–	42.8	-9.2	112	7.89
Little Susitna River at Carl Paulson Place	LS-03	6.1	nd	–	nd	–	110	6.45
Little Susitna River at Moose Meadows Road	LS-04	10.3	159.0	-3.0	62.8	20.0	107	5.76
Little Susitna River at Sushana Road	LS-05	12.0	191.0	32.0	nd	–	nd	nd
Little Susitna River at Schrock Road	LS-06	15.8	219.0	28.0	82.3	19.5	100	5.16
Little Susitna River near Silver Drive	LS-07	22.6	250.0	31.0	111.0	28.7	107	4.96
Little Susitna River near No Name Hill Drive	LS-08	28.6	257.0	7.0	nd	–	nd	nd
Little Susitna River at Parks Highway	LS-09	31.7	279.0	22.0	121.0	10.0	136	1.18

River-water samples collected during the October 2011 seepage run show changes in water chemistry with longitudinal distance along the river (table 6). The chloride concentration of river-water samples gradually decreased from 7.49 mg/L at the most upstream sample point to 1.18 mg/L at the most downstream sample point. The specific conductance of the river water remained relatively constant at 110 μS/cm along most of the study reach before increasing to 136 μS/cm at the most downstream sample point. The gradual dilution of chloride in the river water was not accompanied by a decrease in specific conductance; therefore, other chemical species must have increased to account for the similar levels of dissolved solids indicated by the specific conductance. Assuming the chloride concentrations in the upstream river reaches to be representative of water leaving the Little Susitna River Valley, these data may account for sources of chloride in shallow groundwater downgradient from the Little Susitna River Valley. It is possible that the dilution of chloride along the study segment of the Little Susitna River may be attributed to an inflow of surface water or groundwater with very low chloride concentrations.

Interaction Between Groundwater and Small Streams

Numerous groundwater-connected small streams are found across the Matanuska-Susitna Valley. The largest of these streams in the core area—Lucile Creek, Meadow Creek, Wasilla Creek, and Cottonwood Creek—and their tributaries constitute subwatersheds. Rates of groundwater exchanges with these small streams have not been studied. During 2010–2011, field investigations of groundwater/ surface-water interaction were undertaken along each of the four streams. Data collection included seepage runs for use in gain or loss analysis and collection of water samples from streams and nearby wells; environmental tracers including temperature, specific conductance, chloride, and water isotopes were analyzed.

Lucile Creek

Lucile Creek originates as outflow from Lucile Lake and flows for approximately 13 mi before its confluence with Little Meadow Creek (fig. 17). During the ice-free periods of 2010 and 2011, the net difference in streamflow between the USGS streamgages at Lucile Lake and the confluence with Little Meadow Creek ranged from 0.8 to 3.6 ft³/s; the mean streamflow difference was 2.1 ft³/s (fig. 18). Kikuchi and others (2012) identified a gaining reach of Lucile Creek beginning approximately 5.5 mi downstream from the Lucile Lake outlet within which nearly all the observed gain in streamflow occurs. This characterization was based on seepage and longitudinal changes in water temperature, specific conductance, and isotopic composition from measurements during base-flow conditions. Decreasing specific conductance over the length of the stream was attributed to the influence of water from a regional confined aquifer in local hydraulic contact with a riparian unconfined aquifer. This interpretation was supported by analysis of borehole lithologies from nearby wells. The resulting estimates of streamflow gains to and losses from groundwater are summarized in table 7. Summing the gain or loss estimates and averaging over all four seepage runs yield an estimated value of 1,800 acre-ft/yr for groundwater outflow to Lucile Creek.

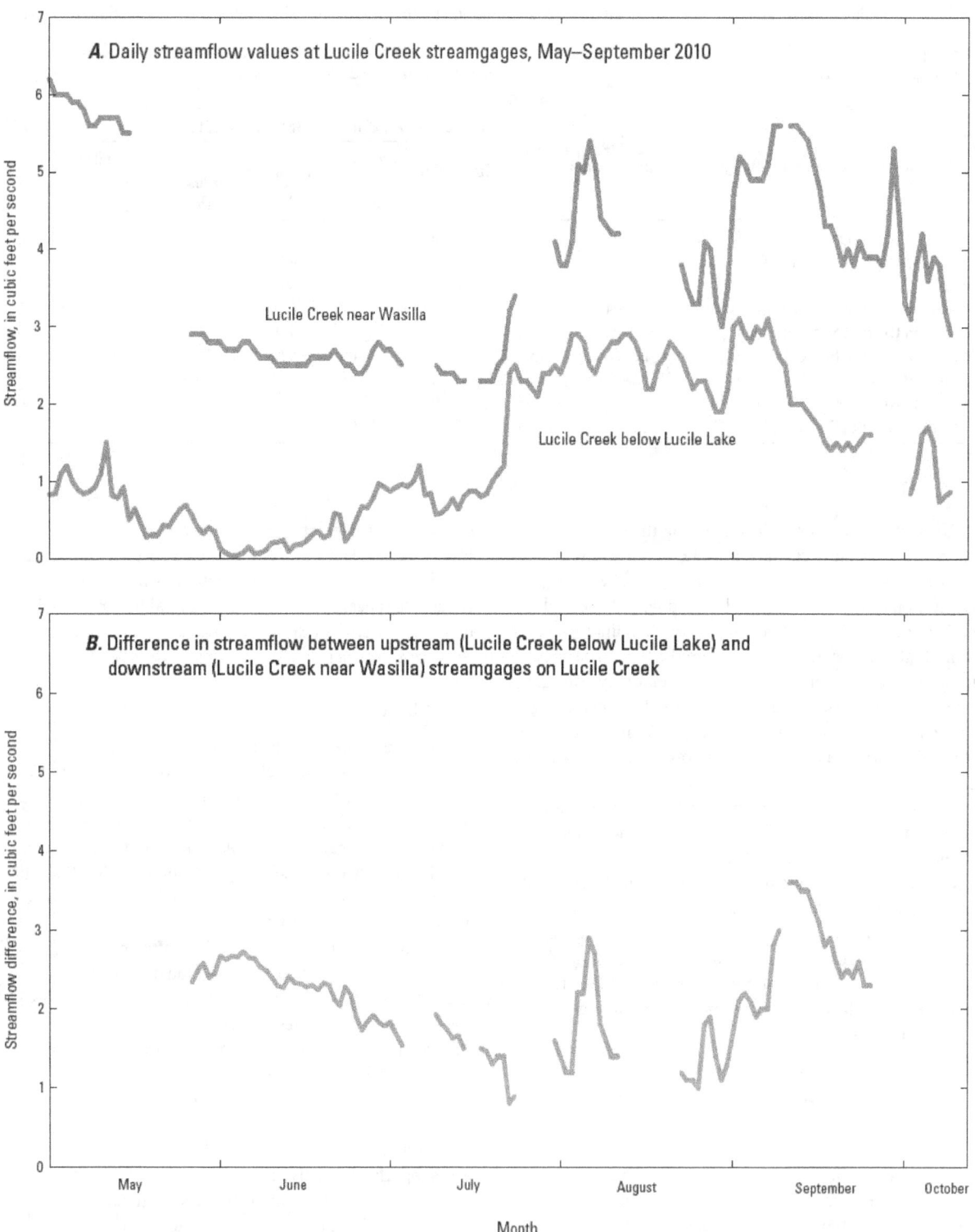

Figure 18.   Change in streamflow along Lucile Creek, Alaska, May 2010–September 2010. (*A*) Daily streamflow values, (*B*) Difference in streamflow between upstream and downstream streamgages.

Table 7. Streamflow measurements for sites along Lucile Creek, Alaska.

[Streamflow measurements are used to determine streamflow gains (positive values) and streamflow losses (negative values).**Abbreviations:** ft³/s, cubic feet per second; nd, no data; –, not calculated]

Measurement site or streamgage	Map identification No.	River mile	October 20, 2009		May 21, 2010		June 29, 2010		September 30, 2010	
			Streamflow (ft³/s)	Gain or loss (ft³/s)	Streamflow (ft³/s)	Gain or loss (ft³/s)	Streamflow (ft³/s)	Gain or loss (ft³/s)	Streamflow (ft³/s)	Gain or loss (ft³/s)
Lucile Creek below Lucile Lake (15286400)	LC-01	0.00	1.98	–	0.42	–	0.92	–	0.48	–
Lucile Creek at Bailey Lane	LC-02	0.41	2.19	0.2	0.32	-0.1	nd	–	nd	–
Lucile Creek at Mack Road	LC-03	1.08	1.87	-0.3	0.49	0.17	0.99	0.07	0.63	0.15
Lucile Creek at Foothills Boulevard	LC-04	2.83	1.66	-0.2	0.59	0.1	0.9	-0.09	0.44	-0.19
Lucile Creek at Vine Road	LC-05	4.14	1.65	0.0	0.41	-0.18	1.06	0.16	0.35	-0.09
Lucile Creek at Sylvan Road	LC-06	5.46	1.68	0.0	1.06	0.65	1.47	0.41	0.78	0.43
Lucile Creek at Misty Lake Road	LC-07	7.10	3.33	1.7	1.65	0.59	2.23	0.76	1.66	0.88
Lucile Creek at Johnson Road	LC-08	9.46	3.8	0.5	3.09	1.44	3.02	0.79	2.72	1.06
Lucile Creek upstream of powerlines	LC-09	9.80	2.96	-0.8	2.94	-0.15	nd	–	nd	–
Lucile Creek near Jeffrey Street	LC-10	11.47	2.7	-0.3	3.39	0.45	nd	–	nd	–
Lucile Creek near Wasilla (15286500)	LC-11	12.87	3.6	0.9	3.56	0.17	3.43	0.41	3.06	0.34

Meadow Creek

Meadow Creek originates from springs in wetlands and small ponds of the Meadow Lakes area (fig. 17); therefore, Meadow Creek acts as a drain for the groundwater system. To investigate groundwater/surface-water interaction along Meadow Creek—between the crossings at Big Lake Road and Beaver Lake Road—field investigations were undertaken during base-flow episodes during summer and early fall 2010. These field investigations included two seepage runs conducted on July 15, 2010, and September 30, 2010, and a longitudinal thermal profile (Vaccaro and Maloy, 2006). During both seepage runs, the streamflow measured at the downstream end of the study reach exceeded the streamflow measured at the upstream end in excess of the measurement error. The streamflow contribution from Lucile Creek accounted for 54 percent of the observed streamflow gain along Meadow Creek (table 8). Two small tributaries draining a lake-wetland complex north of Big Lake were observed, but not measured, between the confluence of Little Meadow Creek with Lucile Creek and the end of the study reach. Therefore, it could not be determined whether the unexplained streamflow gain along Meadow Creek should be attributed to groundwater discharge or to surface-water inflow. The longitudinal thermal profile of Meadow Creek shows a dip in temperature at approximately 3.5 mi corresponding to the confluence with Lucile Creek (fig. 19). Similar dips were not observed between 3.5 and 6 mi, indicating that the water input to Meadow Creek after the confluence with Lucile Creek is from a different, possibly less groundwater-dominated source. However, there is not enough evidence to conclusively identify the source of streamflow gains along this latter reach of Meadow Creek. Subtracting Lucile Creek inflows to Meadow Creek from the streamflow measured at the furthest downstream site and averaging over both seepage runs yield an estimated value of 6,400 acre-ft/yr for groundwater outflow to Meadow Creek.

Table 8. Streamflow measurements on Meadow Creek and tributaries, Alaska.

[Streamflow measurements are used to determine streamflow gains (positive values) and streamflow losses (negative values) between measurement sites Abbreviations: ft³/s, cubic feet per second; –, not calculated; nd, no data]

Measurement site or streamgage	River mile	July 15, 2010		September 30, 2010	
		Streamflow (ft³/s)	Gain or loss (ft³/s)	Streamflow (ft³/s)	Gain or loss (ft³/s)
Little Meadow Creek at Parks Highway	0.0	4.6	–	8.78	–
Little Meadow Creek at Kenlar Road	2.0	4.7	0.1	8.5	-0.28
Lucile Creek near confluence (15286500)[1]	3.0	2.0	–	3.1	–
Little Meadow Creek at Upper Birch Road	3.5	7.0	0.3	14.9	3.3
Meadow Creek off South Lodge Drive	4.9	7.6	0.6	nd	–
Meadow Creek at Beaver Lake Road	5.6	8.3	0.7	14.5	-0.4

[1]Tributary to Little Meadow Creek.

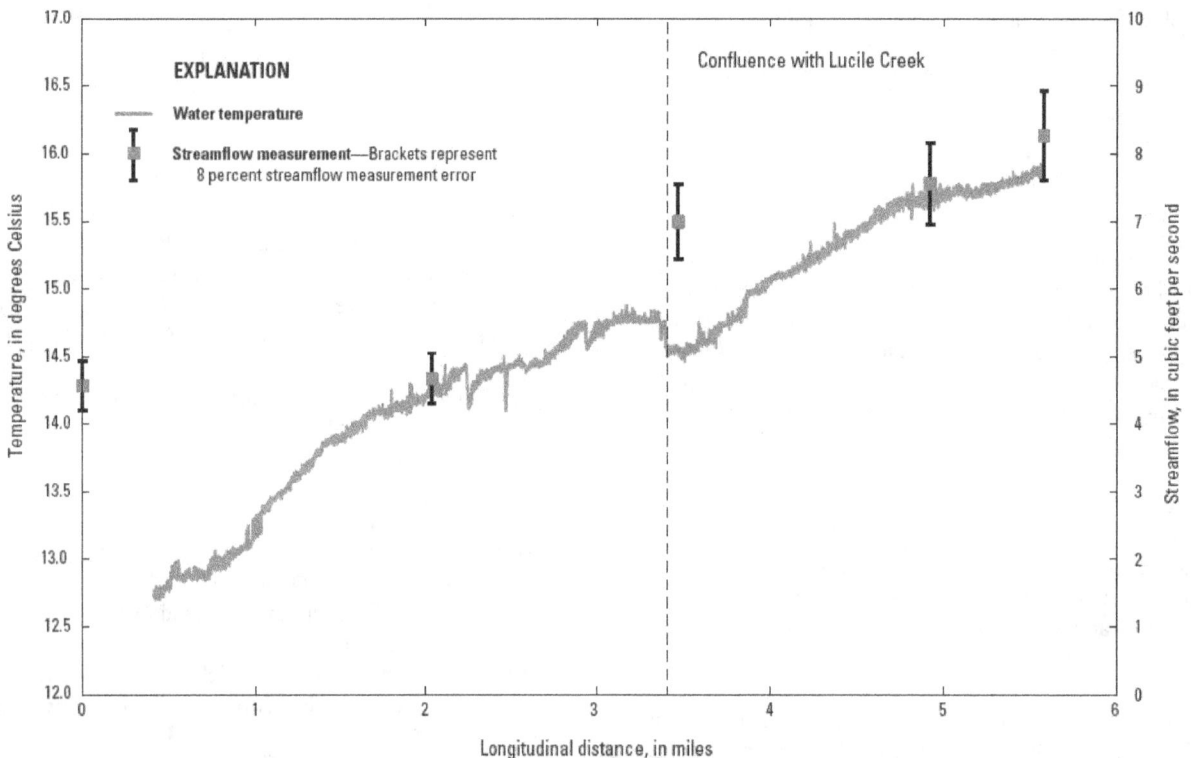

Figure 19. Longitudinal changes in water temperature and streamflow in Meadow Creek, Alaska, July 15, 2010. Distance is measured from Big Lake Road.

Wasilla Creek

The headwaters of Wasilla Creek are north-northeast of Palmer, between Moose Creek and the Little Susitna River (fig. 17). Wasilla Creek receives inputs from two tributaries—Carnegie Creek and an unnamed stream draining Walby Lake —before flowing into Knik Arm. To quantify groundwater discharge to Wasilla Creek, seepage runs were performed during the late summer and fall of 2011. During the late summer (August 8, 2011) seepage run, water samples were collected from the stream and analyzed for water isotopes and chloride concentration. Water isotopes and chloride are commonly treated as chemically inert environmental tracers in studies of groundwater/surface-water interaction. Seepage-run data from both measurement dates (table 9) show streamflow gains that exceed the measurement error between the streamgage at Palmer-Fishhook Road and Bogard Road and between the Parks Highway and Nelson Road. Streamflow losses in excess of the measurement error were detected between Bogard Road and the Parks Highway. The gains in streamflow over the entire study reach were 3.8 and 8.2 ft³/s on the August 8, 2011, and October 18, 2011, seepage runs, respectively. These net gains correspond to 27 and 53 percent of the streamflow measured at the end of the study reach on each respective measurement date.

Table 9.   Streamflow measurements, chloride concentrations, and water isotopic ratios for sites along Wasilla Creek and tributaries, Alaska.

[Streamflow measurements are used to determine streamflow gains (positive values) and streamflow losses (negative values) between measurement sites. **Abbreviations:** ft³/s, cubic feet per second; mg/L, milligrams per liter; VSMOW, Vienna Standard Mean Ocean Water; –, not calculated]

Measurement site or streamgage	Map identification No.	River mile	August 2, 2011		October 12, 2011		August 2, 2011		
			Streamflow (ft³/s)	Gain or loss (ft³/s)	Streamflow (ft³/s)	Gain or loss (ft³/s)	Chloride concentration (mg/L)	Hydrogen-2/ Hydrogen-1 isotopic ratio, per mil relative to VSMOW	Oxygen-18/ Oxygen-16 isotopic ratio, per mil relative to VSMOW
Wasilla Creek at Yarrow Road	WC-01	1.74	7.4	–	10.5	–	0.26	-135.27	-17.57
Wasilla Creek at Palmer-Fishhook Road (15285000)	WC-02	4.61	8.2	0.8	10.0	-0.5	0.70	-135.01	-17.68
Carnegie Creek at Palmer-Fishhook Road[2]	CC-01	5.52	1.1	–	[1]1.3	–	4.96	-132.21	-16.78
Wasilla Creek at Bogard Road	WC-03	7.76	12.5	3.2	12.8	1.5	1.51	-134.66	-17.51
Walby Lake tributary at Trunk Road[2]	WLT-01	8.16	1.2	–	1.3	–	19.20	-128.95	-16.45
Wasilla Creek at Lower Road	WC-04	11.09	12.7	-1.0	10.6	-3.5	4.97	-133	-17.17
Wasilla Creek at Parks Highway	WC-05	12.38	9.9	-2.8	9.9	-0.7	8.30	-132.22	-16.74
Wasilla Creek at Nelson Road	WC-06	14.27	15.6	5.7	14.2	4.3	8.70	-131.51	-16.81

[1] Streamflow measurement from October 21, 2011.

[2] Tributary to Wasilla Creek.

The chloride concentration in Wasilla Creek stream-water samples increased with longitudinal distance in the downstream direction; the greatest change was between the crossings at Bogard Road and the Parks Highway (table 9). A similar pattern was observed for the $\delta^{18}O$ values. These data indicate that groundwater component of streamflow in Wasilla Creek increases the most along an apparently losing reach. Some of this change can be attributed to the input from the Walby Lake tributary along this reach, which showed elevated chloride concentration and higher $^{18}O:^{16}O$ isotopic ratio relative to Wasilla Creek. Chloride concentrations and the $^{18}O:^{16}O$ isotopic ratio in water samples collected from Wasilla Creek at Nelson Road (at the end of the study reach) are very similar to those in a groundwater sample collected from well W111 (USGS site number 614053149134301), located upgradient from Wasilla Creek. Using either chloride or the $^{18}O:^{16}O$ isotopic ratio in a two-component mixing model leads to the conclusion that groundwater constitutes nearly 100 percent of streamflow measured at the Nelson Road site. The groundwater contribution estimated using conservative

tracers is much larger than that estimated by streamflow gain or loss analysis. It is possible that some of the streamflow in Wasilla Creek re-entered bank storage along the downstream reaches and was not measured in the channel at the Nelson Road crossing. Summing the gain or loss estimates and averaging over both seepage runs yield an estimated value of 2,500 acre-ft/yr for groundwater outflow to Wasilla Creek.

Cottonwood Creek

Cottonwood Creek consists of two main segments: a segment connecting Neklason Lake to Cottonwood Lake (upstream segment) and a segment connecting Wasilla Lake to Knik Arm (downstream segment) (fig. 17). On June 1, 2011, a field investigation was conducted on the downstream segment of Cottonwood Creek, including a seepage run and synoptic collection of stream-water samples. Streamflow loss in excess of the 8 percent measurement error (based on the judgment of the field personnel making measurements) was observed between Edlund Road and Marble Road

Table 10.   Streamflow measurements, water temperature, chloride concentration, and water isotopic ratios for sites along Cottonwood Creek and tributaries, Alaska.

[Streamflow measurements are used to determine streamflow gains (positive values) and losses (negative values) between measurement sites. **Abbreviations:** ft³/s, cubic feet per second; mg/L, milligrams per liter; VSMOW, Vienna Standard Mean Ocean Water; –, not calculated]

Measurement site or streamgage	Map identification No.	River mile	June 1, 2011						
			Streamflow (ft³/s)	Gain or loss (ft³/s)	Temperature (degrees Celsius)	Chloride concentration (mg/L)	Hydrogen-2/ Hydrogen-1 isotopic ratio, per mil relative to VSMOW	Oxygen-18/ Oxygen-16 isotopic ratio, per mil relative to VSMOW	
Cottonwood Creek at Matanuska Road	CC-01	0.64	12.33	–	16.07	12.3	-120.43	-14.58	
Cottonwood Creek at Fern Road	CC-02	2.33	11.2	-1.13	15.26	12.6	-120.5	-14.57	
Cottonwood Creek at Edlund Road	CC-03	3.92	12.01	0.81	14.91	14	-120.39	-14.53	
Cottonwood Creek at Suburban Road	CC-04	4.73	10.93	-1.08	14.95	14.2	-120.2	-14.43	
Cottonwood Creek at Marble Road	CC-05	5.77	9.97	-0.96	14.84	14.3	-120.01	-14.34	
Cottonwood Creek Riverdell Road	CC-06	6.47	11.38	1.41	14.63	14.7	-120.29	-14.41	
Cottonwood Creek at Loop Road	CC-07	6.94	10.62	-0.76	13.61	14.7	-119.43	-14.45	
Cottonwood Creek at Surrey Road	CC-08	7.27	11.35	0.73	13.24	14.8	-120.23	-14.7	
Cottonwood Slough at Surrey Road[1]	CS-01	8.3	0.1546	–	9.19	21.9	-125.77	-15.54	
Cottonwood Creek near Hay Flats	CC-09	8.3	11.1	-0.4046	13.58	15.4	-120.27	-14.64	

[1] Tributary to Cottonwood Creek.

(table 10). No streamflow gains in excess of the 8 percent measurement error were observed. The chloride concentration rose steadily with longitudinal distance, and the temperature of the stream water declined from 16.1 to 9.2°C along the entire length of the downstream segment (table 10). The $\delta^{18}O$ value rose from -14.58‰ to -14.34‰ over the first 6 mi of the downstream segment, then declined to -14.75‰ at the end of the downstream segment. The decline in ^{18}O:^{16}O isotopic ratio matches a 1.4°C decrease in water temperature between Riverdell Road and Surrey Road. Similar patterns observed in Lucile Creek were interpreted as evidence of upwelling groundwater from a confined regional aquifer. However, in this case, the decline in water temperature and ^{18}O:^{16}O isotopic ratio was accompanied by an increase in specific conductance and no appreciable increase in streamflow. Furthermore, groundwater samples from wells near Cottonwood Creek have chloride concentrations and ^{18}O:^{16}O isotopic ratios that are much lower than those in the stream water. The results of this field investigation indicate that there was no appreciable groundwater contribution to Cottonwood Creek.

Water Levels

Water-Level Map

Jokela and others (1990) used groundwater-level data from wells in unconfined aquifers and surface-water stage data to construct a water-table map for the Matanuska-Susitna Valley that distinguishes between regional flow directions and local flow directions controlled by local geologic conditions. Moran and Solin (2006) used groundwater-level data from water-table wells and surface-water stage data to construct a similar water-table map. The two maps share some notable similarities. Flow is driven by high water levels in the Little Susitna Valley and then branches into three directions: (1) southeast towards the Matanuska River, (2) directly south into Knik Arm, and (3) west-southwest parallel to the Little Susitna River and towards the Meadow Lakes area. Water-table contours are very steep surrounding Little Meadow Creek, Meadow Creek, and Lucile Creek; these streams appear to be consistently gaining. The water-table contours are less steep around Wasilla Creek and Cottonwood Creek; reaches of these streams appear to be alternately gaining and losing. Finally, there are two pronounced groundwater mounds in the study area; one long mound trends west-southwest between the Little Susitna River and the Lucile Creek subwatershed and includes the Meadow Lakes area. The second groundwater mound trends west-southwest directly south of Lucile Creek and is roughly equidistant between Lucile Lake and Big Lake. Both groundwater mounds correspond to topographic high points in the hummocky terrain.

The groundwater-level data used in the construction of both maps were obtained primarily from driller's logs, with water levels measured at or shortly after the time of well construction; thus, there is some uncertainty as to whether equilibrium had been reached between the well and surrounding aquifer. The locations of wells used in the construction of both maps were approximate and subject to some uncertainty. Furthermore, the period of record for water levels spanned from 1935 to the present, so the water-table maps constructed from these data likely are affected by seasonal and interannual water-level variations at local and subregional scales and do not constitute a snapshot of the water-table elevations. To provide more precise water-level data in conceptualization of the regional aquifer system, a synoptic water-level measurement campaign was undertaken during a 3-week period in summer 2009. Static water levels were manually measured in 135 water wells, and the locations of borehole well casings were measured using Real Time Kinematic (RTK) GPS survey techniques with an accuracy of +0.1 ft. Most of the lakes in the Matanuska-Susitna Valley are hydraulically connected to the groundwater system; therefore, water-surface elevations in 38 lakes were also measured during summer 2009 for inclusion in the synoptic water-level dataset using RTK GPS survey techniques. The locations of wells and lakes included in the synoptic water-level measurement are displayed in figure 20.

Of the 135 wells included in the 2009 synoptic water-level measurement, 12 were completed in bedrock. The remaining wells were completed in unconsolidated sediments. For these wells, physical information on well construction was taken from well driller's logs and was used to classify the water-bearing formation that each well penetrates as either confined or unconfined (appendix D). Wells with a layer of clay, silt, or glacial till above the well opening were considered to have a confining layer present; those wells for which the static water level was above the bottom of the confining layer were classified as confined. Thirty-four of the 135 wells measured in this study are completed in confined aquifers, 46 in unconfined aquifers, and 55 of the wells could not be classified because of a lack of sufficient well-construction information. Synoptic water-level measurement data from lakes and wells completed in unconfined aquifers were synthesized with previous water-table maps of Jokela and others (1990) and Moran and Solin (2006) to construct an updated water-table map (fig. 21). Water levels in open-end wells represent the water table only when vertical hydraulic gradients are negligible. In constructing the updated water-table map, vertical hydraulic gradients were assumed to be negligible. Available data were insufficient to construct water-level contour maps in the lower aquifer units (Lower Permeable Sediments and Tertiary Sedimentary Rocks); however, water-level data for individual wells in those units were still used to calibrate and evaluate the numerical groundwater flow model.

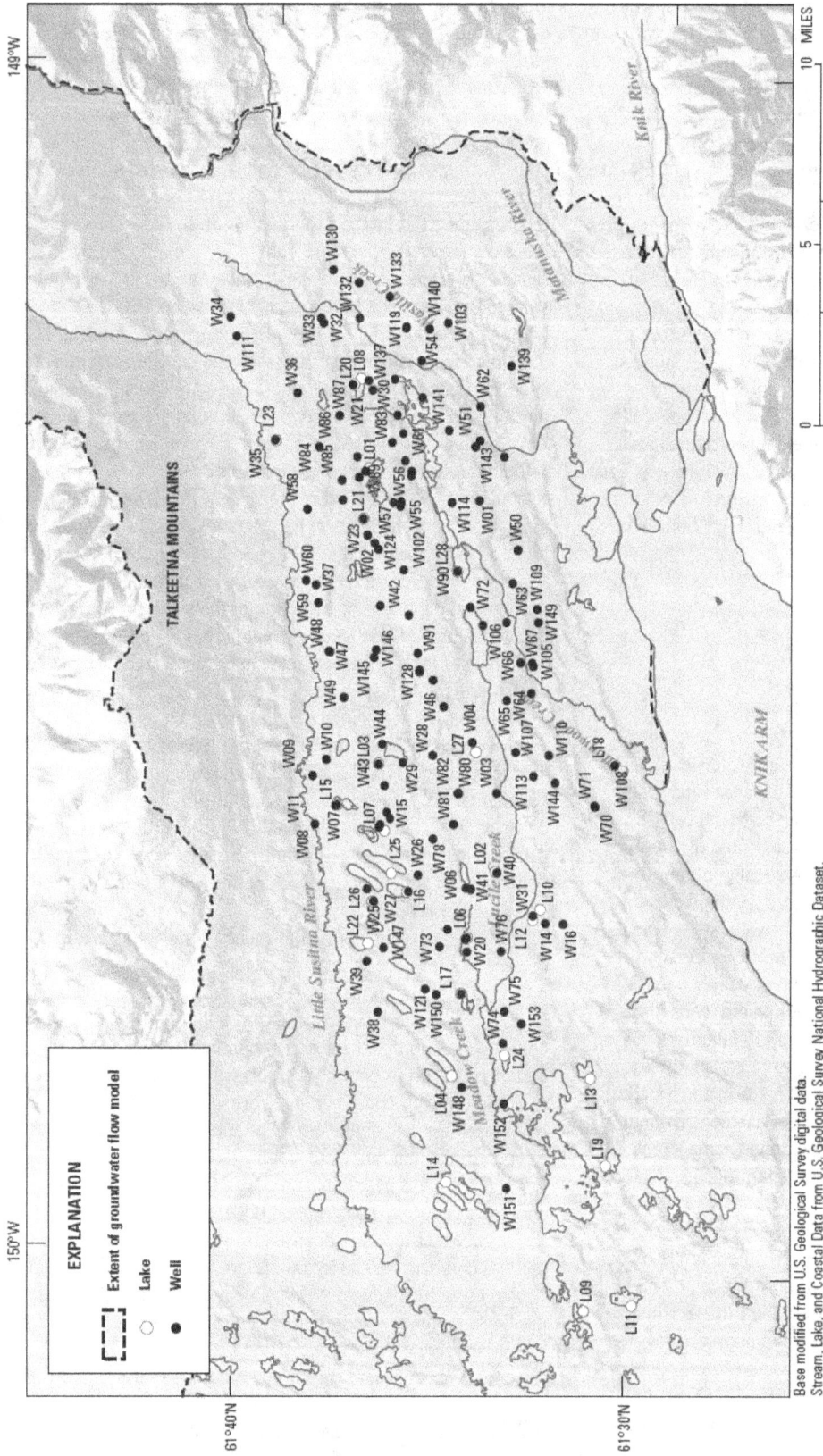

Figure 20.   Location of wells and lakes included in a synoptic water-level measurement, Matanuska-Susitna Valley, Alaska, 2009.

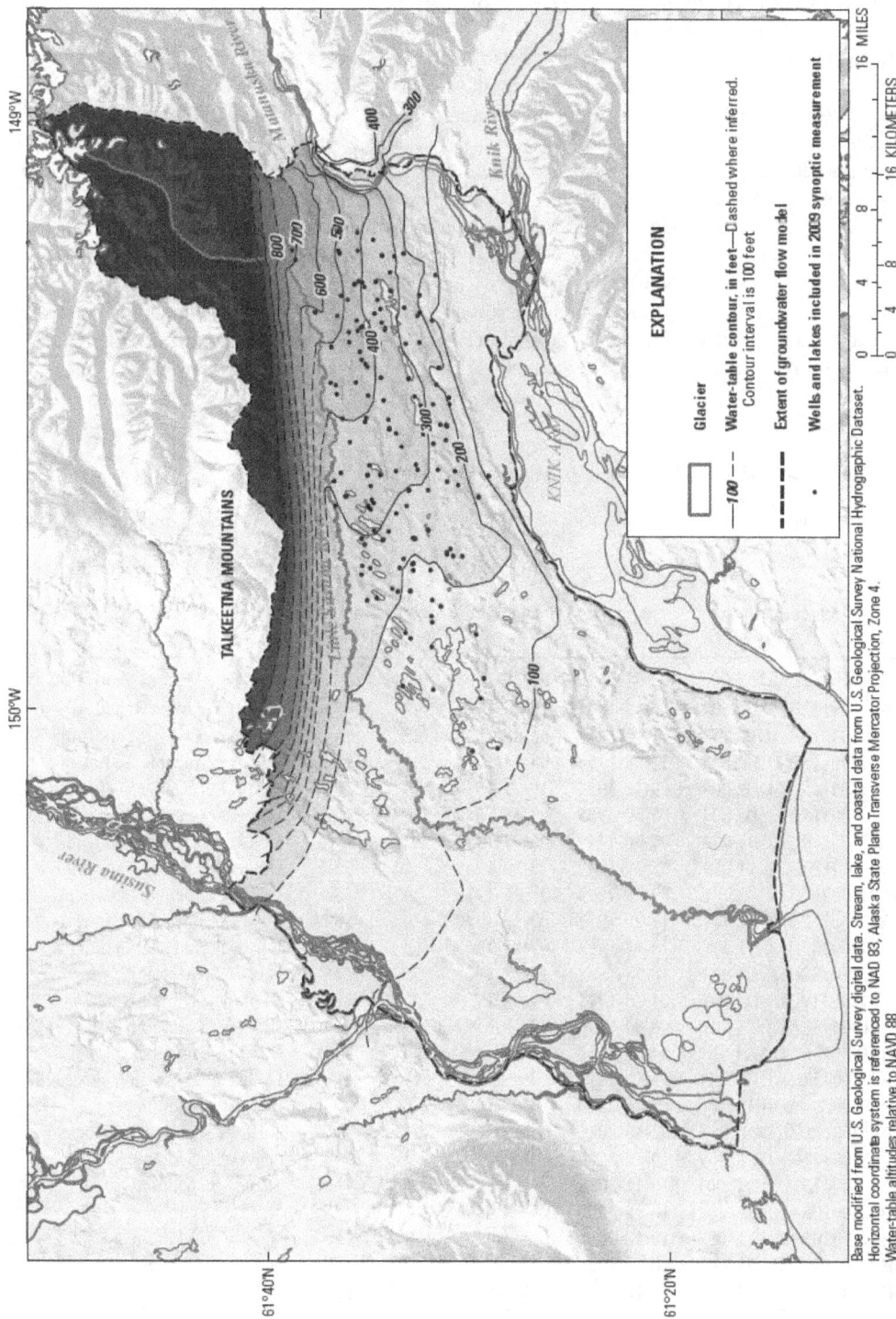

Base modified from U.S. Geological Survey digital data. Stream, lake, and coastal data from U.S. Geological Survey National Hydrographic Dataset.
Horizontal coordinate system is referenced to NAD 83, Alaska State Plane Transverse Mercator Projection, Zone 4.
Water-table altitudes relative to NAVD 88

Figure 21.   Updated water-table contour map and locations of 2009 synoptic water-level measurements, Matanuska-Susitna Valley, Alaska.

Historical time-series groundwater-level data from the Matanuska-Susitna Valley are available through the NWIS database from 1949 to 1993 (table 11). From the 135 wells included in the 2009 synoptic water-level measurement, a subset of 20 wells was selected for continued monitoring during 2010 and 2011; water levels in these wells were manually measured one or two times per year. In addition, pressure transducers were installed in nine wells to automatically record water levels once an hour. Synoptic and continuously monitored water levels were stored in the NWIS database. Nine of the 38 lakes were instrumented with pressure transducers during the ice-free period to record water levels every 15 minutes. Wells and lakes instrumented during this study and wells with historical groundwater data are displayed in figure 22.

Groundwater-Level Fluctuation

Historical water-level records from wells listed in table 11 show considerable temporal variability at seasonal to interannual time scales. During the 1940s and 1950s, groundwater withdrawal rates in the Matanuska-Susitna Valley were much lower than at present; therefore, these records may be considered representative of pre-development conditions. An understanding of the natural hydrogeologic processes responsible for such fluctuation is needed in order to correctly distinguish between naturally occurring and human-induced changes in groundwater levels. The most important controlling factors on natural groundwater-level fluctuation are climate, hydrogeologic setting, relation to surface-water features, and well depth.

Table 11. Physical data for wells with time series water level records, Matanuska-Susitna Valley, Alaska.

[Abbreviations: USGS, U.S. Geological Survey; nd, no data]

Well No.	USGS site No.	USGS local well No.	Begin date	End date	Hole depth (feet)	Elevation datum (feet)	Hydrogeologic unit
W157	613341149144001	SA01700115DBDA1 006	07-31-1949	07-29-1964	108	[1]138	Holocene Outwash and Alluvium
W158	613403149151001	SA01700115BDAB1 003	06-14-1955	10-22-1993	40	[1]173	Holocene Outwash and Alluvium
W159	613406149070001	SA01700217AADD1 004	11-15-1955	06-28-1966	282	[1]164	Holocene Outwash and Alluvium
W160	613406149152101	SA01700115BACD1 004	11-18-1954	08-14-1974	295	[1]172	Lower Permeable Sediments
W161	613406149152102	SA01700115BACD2 004	09-12-1955	10-22-1993	259	[1]160	Lower Permeable Sediments
W162	613406149152103	SA01700115BACD3 004	09-10-1955	08-15-1974	313	[1]172	Lower Permeable Sediments
W163	613417149065401	SA01700209CCCB1 009	07-30-1949	06-01-1991	83	[1]170	Holocene Outwash and Alluvium
W164	613455149263601	SB01700110BCAA1 025	08-07-1949	12-30-1969	28	[1]350	Holocene Outwash and Alluvium
W165	613630149084301	SA01800232BCCB1 001	09-29-1958	01-25-1971	624	[1]388	Lower Permeable Sediments
W166	613630149084302	SA01800232BCCB2 001	12-30-1952	02-26-1980	170	[1]388	Lower Permeable Sediments
W167	613630149084303	SA01800232BCCB3 001	10-11-1952	07-03-1969	165	[1]375	Lower Permeable Sediments
W168	613634149081301	SA01800232BDBC1 002	10-14-1952	12-23-1965	79	[1]365	Naptowne Moraine
W169	613804149092901	SA01800219DBBD1 007	07-30-1949	09-02-1954	58	[1]488	Naptowne Moraine
W170	614147150013801	SB01900432ADBD1 001	08-04-1976	09-20-1993	69	[1]260	Naptowne Moraine
W55	613634149214401	SB01800136ADAA1	04-23-2010	Present	100	[2]448.5	Naptowne Moraine
W172	613246149475901	SB01700322CADC1 011	04-15-2011	Present	13	nd	Holocene Outwash and Alluvium
W71	613122149363701	SB01700234ACBB1	02-24-2010	Present	120	[2]293.0	Naptowne Moraine
W37	613836149265901	SB01800115CCCC1 004	02-18-2011	Present	200	[2]488.9	Tertiary Sedimentary Rocks
W175	613527149464301	SB01700302CABC1 007	10-18-2008	Present	4.9	[2]229.3	Naptowne Moraine
W176	613343149542901	SB01700413CAAD1 001	10-11-2008	Present	3.1	[2]169.6	Naptowne Moraine
W177	612955149520301	SB01600308ABBB1 002	10-18-2008	Present	4.3	[2]120.2	Naptowne Moraine
W178	613018149514701	SB01600305AABC1 002	10-18-2008	Present	4.3	[2]130.8	Naptowne Moraine
W179	613337149513001	SB01700317CCBD1 001	10-11-2008	Present	7.4	[2]151.6	Holocene Outwash and Alluvium
W180	613344149513601	SB01700317DAAC1 002	10-18-2008	Present	9.3	[2]154.6	Holocene Outwash and Alluvium

[1]Datum is land-surface elevation.

[2]Datum is top of well casing.

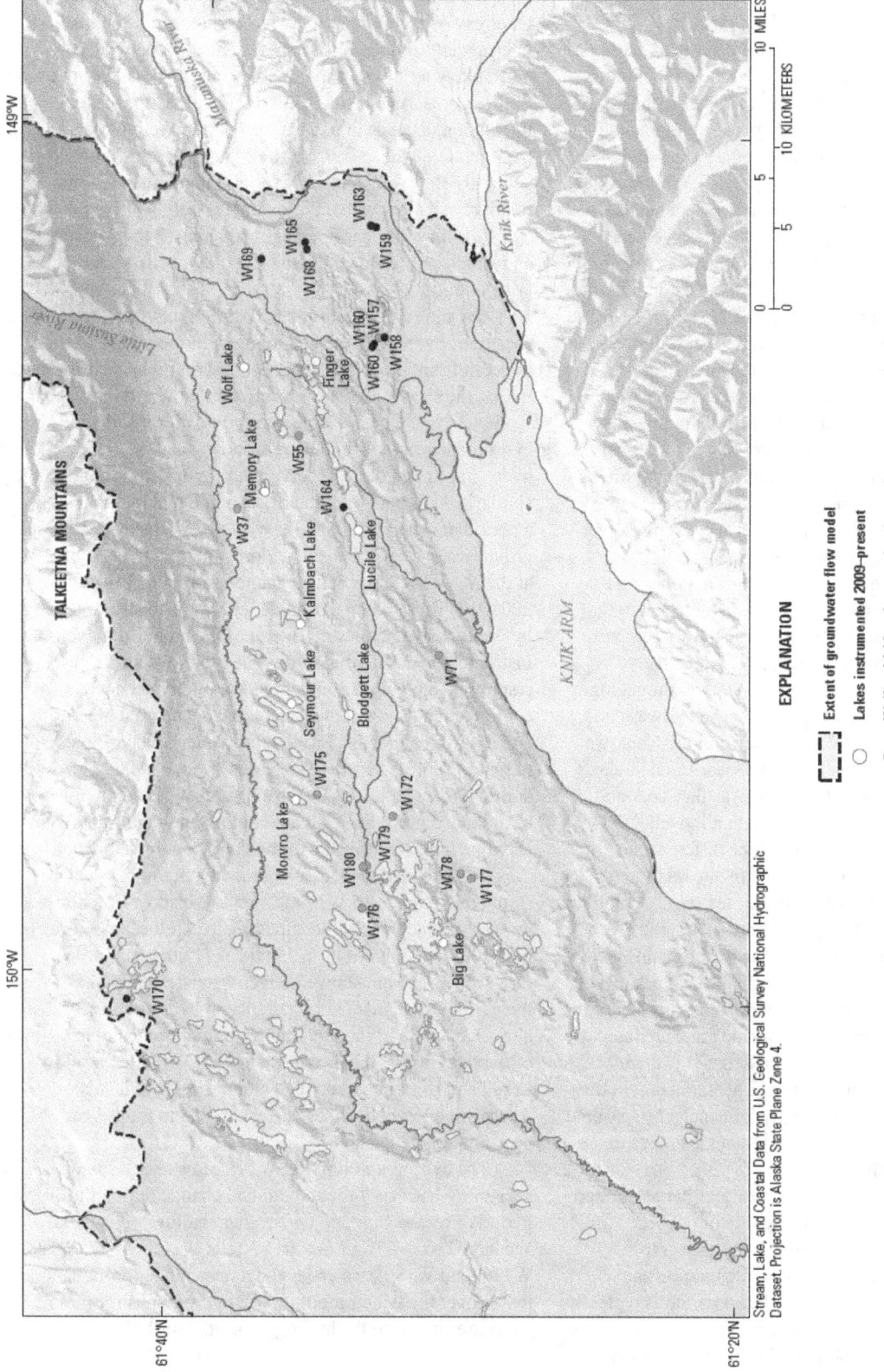

Figure 22. Locations of lakes and wells instrumented during this study and wells with historic groundwater-level data, Matanuska-Susitna Valley, Alaska.

On the basis of observations of 1949–1955 groundwater-level and precipitation data, Trainer (1960) hypothesized that deep wetting of soils and surficial geologic deposits during previous years creates favorable conditions for deep percolation and ultimately groundwater recharge. In this conceptual framework, variability between high and low groundwater levels may be explained by the succession of several years with above-average precipitation. Groundwater levels in wells W164 and W169 illustrate the dependence of groundwater levels on precipitation in successive years (fig. 23A). From 1917 to 2011, the mean annual precipitation measured at the MAES weather station was 15.25 in. During 1950, the annual precipitation was 7.3 in. (fig. 23C); groundwater-level declines ranged from 3 to 10 ft during this year. Rising groundwater levels during the late fall and winter of 1951–52 coincide with a return to average precipitation during July–September. The water-level rise in both wells is of greater magnitude during the winter of 1952–53 than the previous winter, despite similar summer precipitation in 1952 and 1953. This difference may be explained by the dependence of groundwater recharge on a succession of wet years, as hypothesized by Trainer (1960).

Groundwater-level fluctuation depends not only on the timing of precipitation but also on hydrogeologic setting. For example, wells W164 and W169 are shallow open-end wells with opening depths of 28 and 58 ft, respectively; however, each well is completed in a different hydrogeologic unit. Well W164 is completed in the Holocene Outwash and Alluvium unit, and well W169 is completed in the Naptowne Moraine unit. Furthermore, well W164 is near Lucile Lake, whereas well W169 is on a hill between Wasilla Creek and the Matanuska River. Water-level fluctuation in well W164 is less pronounced than in well W169. The difference in these two hydrographs can be explained by the relation of the well to topographic and surface-water features. Hydraulic connection to Lucile Lake likely moderates interannual water-level fluctuation in well W164; furthermore, this well is in a drainage basin formed by the ancient Matanuska River and likely collects water from adjacent uplands throughout the year. In contrast, well W169 is in an upland area, upgradient from Wasilla Creek and the Matanuska River. It is therefore likely that groundwater in the Naptowne Moraine unit in the vicinity of well W169 is more strongly influenced by seasonal in-place recharge variability, with continuous drainage to nearby surface-water features.

The slow movement and storage of water in unsaturated sediments dampens the seasonal variability of deep percolation and groundwater recharge with respect to the seasonal variability of the inputs—precipitation and snowmelt. Furthermore, hummocky glacial terrain is typically characterized by local, intermediate, and regional groundwater flow systems (Winter and others, 1998); local flow systems typically are connected hydraulically to surface-water features and are more sensitive to climate variation than are intermediate and regional flow systems. Deeper wells are more likely to intersect intermediate and regional groundwater flow systems. As a result, the magnitude of water-level fluctuation depends in part on well depth. Well hydrographs from relatively deep wells in the Matanuska-Susitna Valley generally show less temporal variability than those from relatively shallow wells. For example, wells W157 and W163 (fig. 23B) have deeper openings (108 and 83 ft, respectively) and smaller water-level fluctuations than wells W164 and W169 during 1949–54.

The nine wells instrumented during this study were in a variety of hydrogeologic and topographic settings and ranged in depth from 3 to 200 ft. The well hydrographs recorded at five of the sites established during this study are shown in figures 24A-C. Considerable variability was observed in well hydrographs from different areas; however, historical well hydrographs and those observed during this study share some similarities. In particular, below-average precipitation over a period of several years led to a relatively steady decline in groundwater levels. The annual precipitation deficit observed at the MAES weather station ranged from 2 to 3 in. during 2008–10. These precipitation deficits were accompanied by water-level declines in several of the monitoring wells instrumented during this study. The groundwater-level responses to precipitation deficits varied between wells completed in different hydrogeologic or topographic settings and at different depths. Pronounced water-level declines ranging from 0.5 to 1 ft/yr were observed in wells W55, W71, and W179 (fig. 24A-B). Wells W55 and W71 are on hills of mounded glacial till and are completed in the Naptowne Moraine unit; well W55 is immediately upgradient from Wasilla Lake, and well W71 is immediately upgradient from Knik Arm. Well W179 is next to Meadow Creek in a topographic depression occupied by the Holocene Outwash and Alluvium unit. Wells W55 and W71 are much deeper than well W179 but showed larger water-level declines over the period of record. This difference in hydrograph response may be because of the topographic and hydrogeologic settings of the three wells; the proximity of well W179 to Meadow Creek and Big Lake likely had a moderating influence on the water-level declines corresponding to multi-year precipitation deficits.

The hydrograph for well W175 shows large seasonal variability but very little interannual change during the 3-year period of record (fig. 24B). For other shallow wells in the Meadow Lakes—Big Lake area (wells W176, W177, W178, W179, and W180), water-level declines ranged from 0.5 to 3 ft over the winter months. The hydrographs for these wells are similar to the hydrograph for well W179 (fig. 24B).

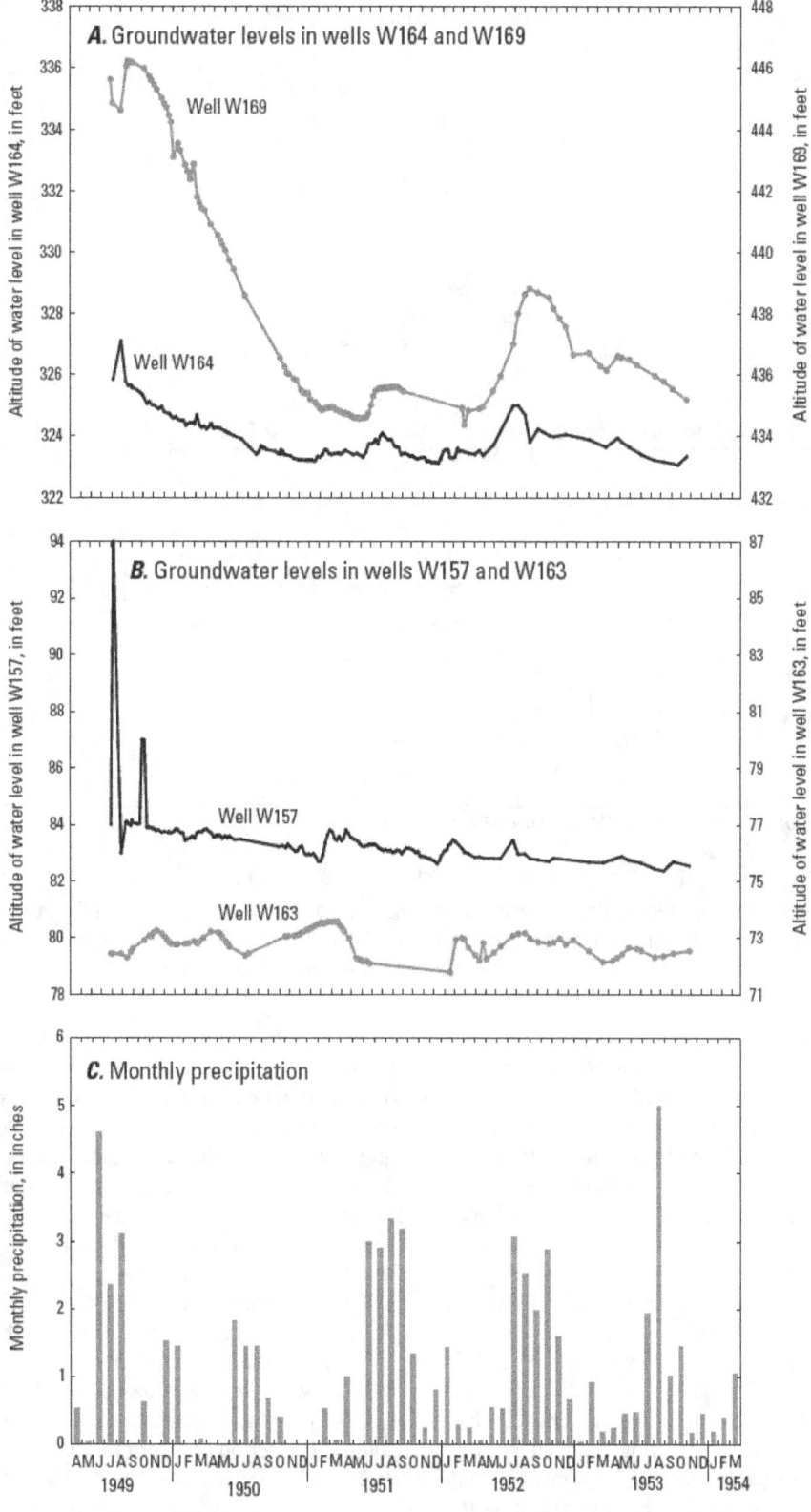

Figure 23. Monthly precipitation and groundwater levels, Matanuska-Susitna Valley , Alaska, 1949–1954. (*A*) Groundwater levels in wells W164 and W196; (*B*) Groundwater levels in wells W157 and W163; (*C*) Monthly precipitation measured at the Matanuska Agricultural Experimental Station near Palmer, Alaska.

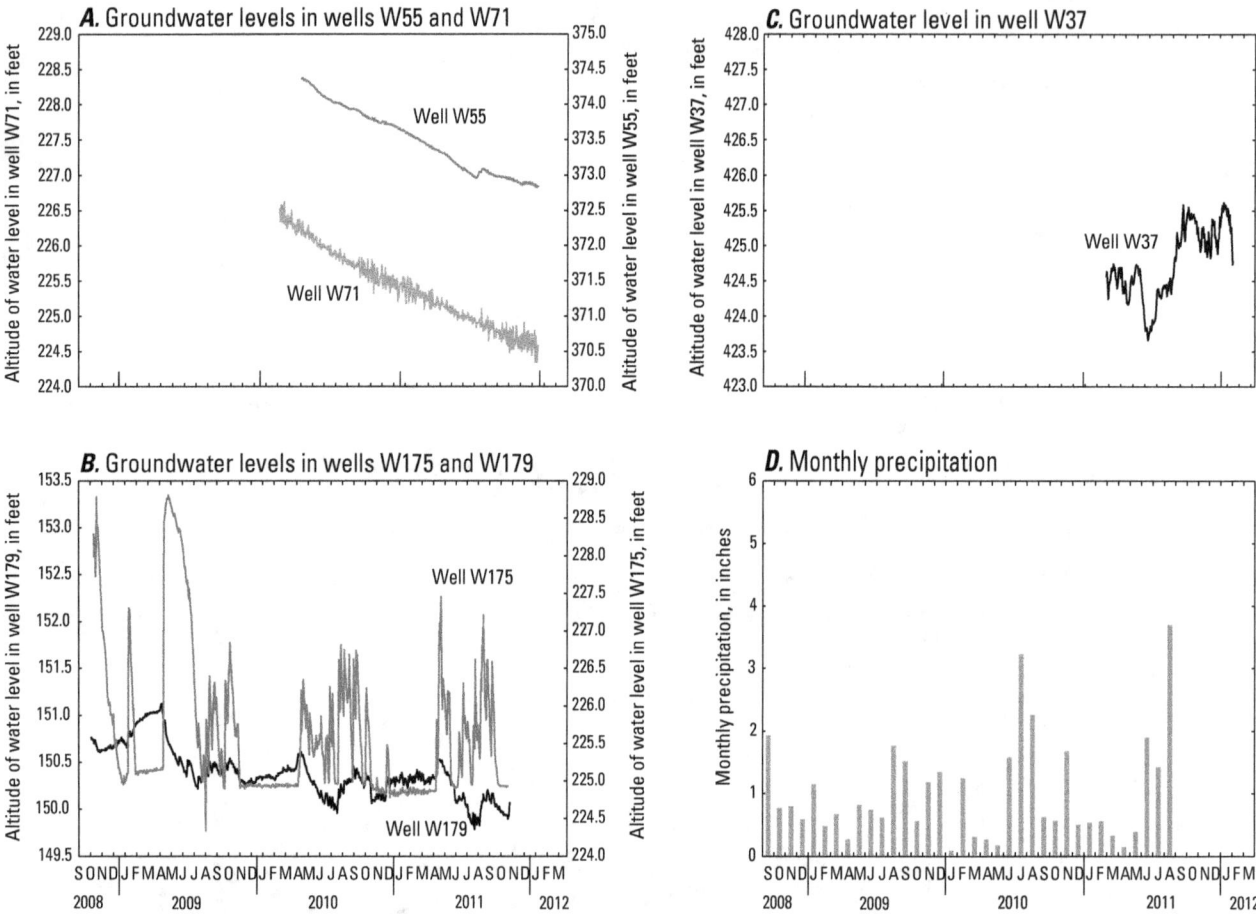

Figure 24. Monthly precipitation and groundwater levels, Matanuska-Susitna Valley, Alaska, 2008–2012. (*A*) Groundwater levels in wells W55 and W71; (*B*) Groundwater levels in wells W175 and W179; (*C*) Groundwater levels in well W37; (*D*) Monthly precipitation measured at the Matanuska Agricultural Experimental Station near Palmer, Alaska.

In comparison, the water level in well W175 remained relatively flat over the winter months. The difference in the hydrographs for well W175 and wells W176, W177, W178, W179, and W180 may be explained by the unique topographic and hydrogeologic setting of well W175. This well is in a closed topographic depression in the Meadow Lakes area, east of Morvro Lake, and is relatively shallow (4.9 ft). No surface-water features drain the aquifer in this particular location, and the constant water levels during the winter indicate negligible leakage to deeper groundwater. It is therefore likely that well W175 is representative of perched aquifers throughout the Meadow Lakes area. Existing hydrogeologic and vegetation data support this interpretation. Gracz (2009) characterized the hydromorphic setting in this area as ripple trough peatlands supporting bogs and lakes, also noting that the presence of bogs indicates precipitation, rather than groundwater discharge, is the predominant water input source. The surficial unconsolidated sediments in this area are identified as glacial and lacustrine sediments (Wilson and others, 2009); hydrogeologic sections drawn

through the Meadow Lakes area as part of this study show the presence of diamict and fine sediments at depth. Comparison of the hydrographs from well W175 with well W179 demonstrates the controlling influence of hydrogeology on groundwater-level fluctuation; the absence of surface-water drainage features and the presence of relatively impermeable sediments at depth mediates the influence of multi-year precipitation deficit on groundwater levels.

Well W37 is finished in the Tertiary Sedimentary Rock unit and penetrates a sequence of interbedded shale and coal. As is common in other wells finished in this hydrogeologic unit, numerous coal seams within the open interval of well W37 are water bearing. Groundwater-level fluctuations in well W37 are therefore representative of a fractured-rock aquifer. The period of record for this well is insufficient to determine any interannual trends. Water-level fluctuations from March 2010 through February 2011 may have responded to seasonal variability in precipitation; rising groundwater levels corresponded to the arrival of late summer rains.

Additional water-level observations in this well may elucidate flow processes in the Tertiary Sedimentary Rock unit, including flow between this hydrogeologic unit and overlying unconsolidated sediments.

Lake-Stage Fluctuation

Stage variability among the nine lakes monitored during this study corresponds to interannual, seasonal, and daily time scales (fig. 25A-E). Three distinct interannual patterns were observed in lake stages during the ice-free periods of 2009–11. For Big Lake, Blodgett Lake, and Lucile Lake, the mean annual lake stage remained relatively constant during the period of record. For Wolf Lake, Memory Lake, and Morvro Lake, declines of 1–2 ft were measured during the period of record. For Kalmbach Lake, Seymour Lake, and Finger Lake, the mean lake stage increased between the 2009 and 2010 ice-free seasons then remained constant through the 2011 ice-free season. Seasonal stage variability in all the monitored lakes followed a characteristic pattern over the course of the ice-free season. Lake stage is relatively high early in the season following snowmelt and thawing of the frozen soil then steadily declines until middle to late July. At this time, the lake stage rises in response to the arrival of late summer rains then declines following the end of the rainy period. At daily time scales, the lake stages responded almost immediately to strong storms. Precipitation is spatially variable over the Matanuska-Susitna Valley, so daily fluctuation in lake stage due to storms differs between the different monitored lakes.

Comparing over-winter (ice-affected) water balances with over-summer (ice-free) water balances facilitates grouping of lakes and the conceptualization of lake-groundwater interactions. Over-winter and over-summer changes in lake stage were computed for all monitored lakes. When mean lake stage changes during ice-affected and ice-free periods are plotted, the lakes form clusters in different quadrants of the plot consistent with qualitative observations of interannual changes in lake stage (fig. 26). For lakes that plot in quadrant II, stage rose over the ice-affected period but declined over the ice-free period. For lakes that plot in quadrant III, stage declined over the ice-affected period and over the ice-free period. For lakes that plot in quadrant IV, stage declined over the ice-affected period, but rose over the ice-free period.

This analysis of lake-stage change over ice-affected and ice-free periods was synthesized with a lake classification scheme described by Jokela and others (1990) to provide a conceptual framework of lake-groundwater interactions in the study area. Snowmelt and under-ice drainage to outlet streams and groundwater are the dominant processes affecting lake stage over the ice-affected period. Rain, inflows from inlet streams and groundwater, outflows to outlet streams and groundwater, and losses from evapotranspiration are the dominant processes affecting lake stage over the ice-free period. For lakes that plot in quadrant II, inflows from snowmelt exceed over-winter outflows. During the ice-free period, outflows and evapotranspiration exceed rain and inflows from surface water and groundwater. For lakes that plot in quadrant III, over-winter outflows exceed snowmelt. During the ice-free period, outflows and evapotranspiration exceed rain and inflows from surface water and groundwater. For lakes that plot in quadrant IV, over-winter outflows exceed snowmelt. During the ice-free period, rain and inflows from surface water and groundwater exceed outflows and evapotranspiration.

Jokela and others (1990) classified lakes in the Matanuska-Susitna Valley on the basis of the presence of surface-water features. Drainage lakes have inlets and outlets, seepage lakes have no inlet or outlet, inlet lakes have only inlets, and outlet lakes have only outlets. This classification scheme aids interpretation of the over-winter and over-summer water balances illustrated in figure 26.

Finger Lake, Kalmbach Lake, and Seymour Lake fall in quadrant II; Finger Lake and Kalmbach Lake are seepage lakes, and Seymour Lake is an outlet lake. The water table is locally mounded surrounding the three lakes, and lake hydrographs from 2009 to 2011 indicate that the stage in these three lakes declined quickly during late June and early July. These data suggest that the over-winter rise in lake stage resulted from local inflows from surface water and groundwater associated with snowmelt and thawing of the soil. Seasonal inflows to these lakes following break-up and corresponding lake-stage increases reflect local-scale, rather than regional-scale, processes in lake-groundwater systems.

Memory Lake and Morvro Lake fall in quadrant III; both are seepage lakes but are in different topographic settings. Memory Lake is at a topographic high point, and Morvro Lake is in a wetland complex north of Big Lake. In the absence of outlet streams, the relative over-winter decline in lake stage indicates steady outflow to groundwater. For Morvro Lake, the decline in lake stage is much higher during the ice-free period than during the winter period. It is likely that evapotranspiration from surrounding wetland vegetation accounts for the relatively large declines during the study period. The steady declines in Memory Lake, coupled with its upgradient position, indicate that Memory Lake is a regionally important feature that recharges the groundwater system. For Morvro Lake, summer lake-stage decline attributed to evapotranspiration exceeds over-winter lake-stage decline by a factor of eight, indicating negligible importance in the regional flow system.

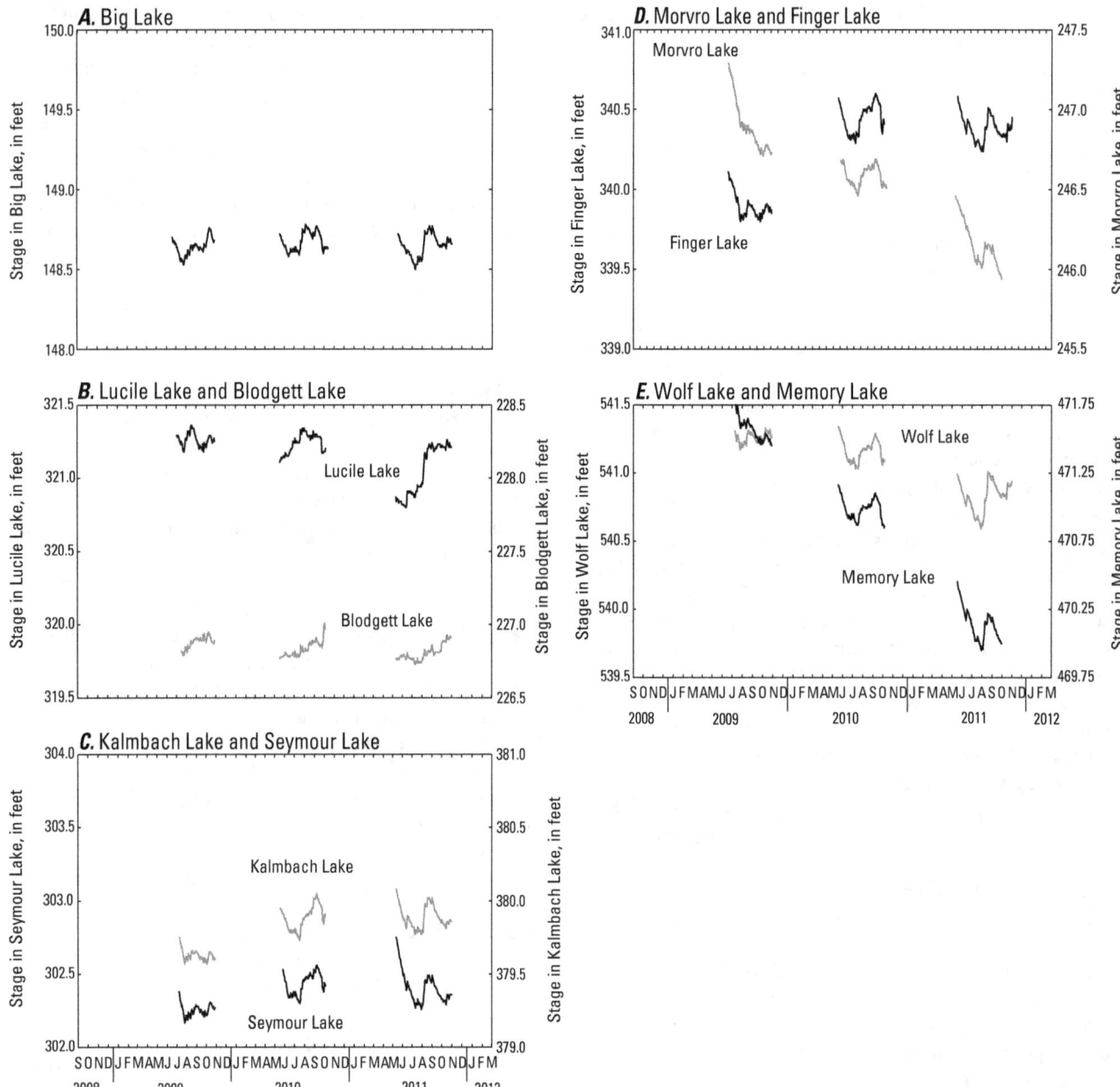

Figure 25. Lake stages measured during ice-free conditions, Matanuska-Susitna Valley, Alaska, 2009–2011. (*A*) Big Lake; (*B*) Lucile Lake and Blodgett Lake; (*C*) Kalmbach Lake and Seymour Lake; (*D*) Morvro Lake and Finger Lake; (*E*) Wolf Lake and Memory Lake.

Lucile Lake and Blodgett Lake fall in quadrant IV; both are outlet lakes and are in similar topographic settings. The average outflow from Lucile Lake, measured at the nearby USGS streamgage on Lucile Creek, was 1.4 ft³/s during the ice-free seasons of 2009–10. The stage of Lucile Lake is regulated by a weir at the outlet to Lucile Creek. In contrast to the lakes plotting in quadrant II, a steep decline in lake stage was not observed in Lucile Lake or Blodgett Lake early in the ice-free period; snowmelt input is routed into their respective outlet streams, so that the lake stage does not rise during snowmelt and thawing of the soil. Lake-stage declines during the winter period because water leaves these lakes as under-ice drainage in excess of groundwater inflow. Lake stage is relatively constant from break-up to the onset of late summer rains, indicating that groundwater inflows are roughly equal to the surface-water outflows and evapotranspiration. Similar to other lakes, the stage in both Lucile Lake and Blodgett Lake rose steeply with the onset of late summer rains. Steady groundwater outflow during the winter period and groundwater inflow during late spring and early summer, inferred from lake hydrographs, indicate that Lucile Lake and Blodgett Lake are important features in the regional groundwater flow system.

In figure 26, Big Lake and Wolf Lake have no strong association with any of the four quadrants. Big Lake is

a drainage lake, and water-table maps indicate that Big Lake also is embedded in the regional flow system. During 2009–11, seasonal variability of stage in Big Lake was similar to other lakes, but there was little interannual variability. Wolf Lake is a seepage lake upgradient from Memory Lake, and water-table maps indicate regional groundwater flow through Wolf Lake. In addition to seasonal variability of lake stage similar to the other monitored lakes, an interannual trend in declining stage was observed for Wolf Lake. Stage variability in Wolf Lake closely tracks stage variability in Memory Lake (fig. 25E).

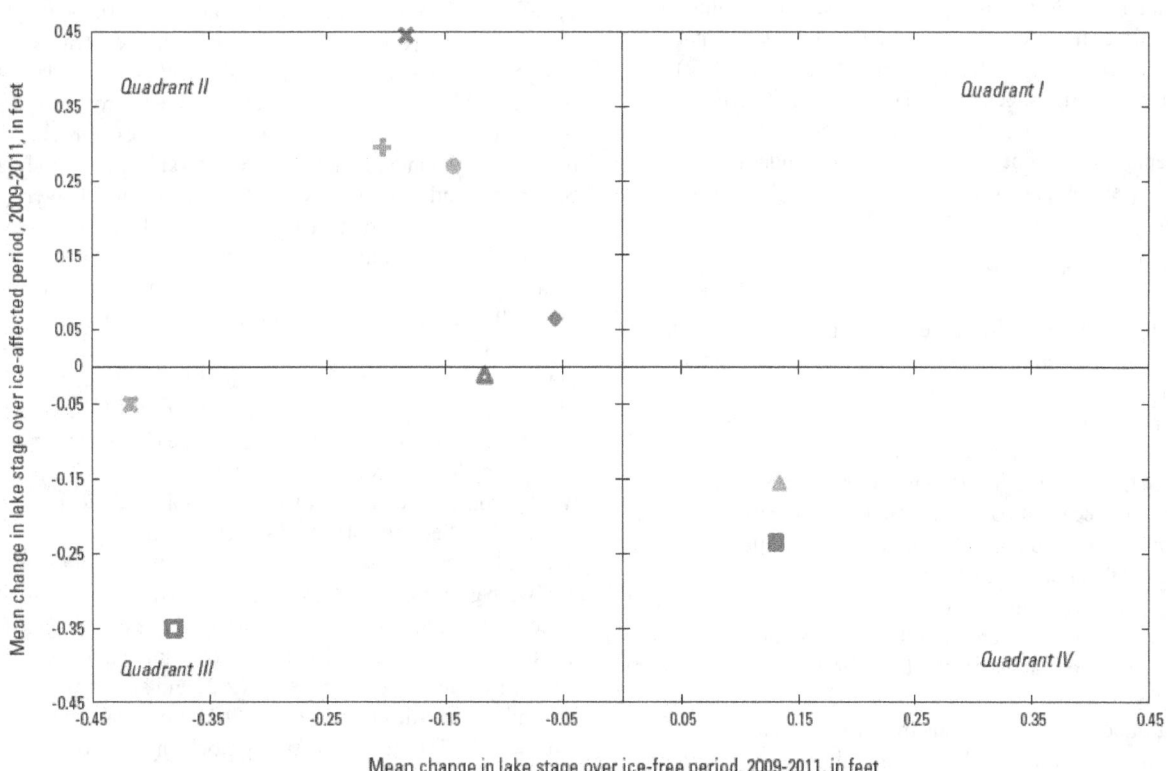

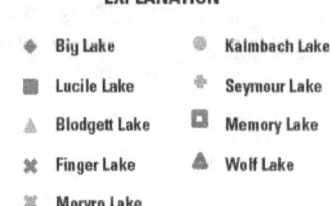

Figure 26.   Lake-stage changes for monitored lakes over ice-affected and ice-free periods, Matanuska-Susitna Valley, Alaska, 2009–2011.

Groundwater Flow Model

Groundwater flow was simulated using MODFLOW-NWT (Niswonger and others, 2011). MODFLOW-NWT numerically solves the three-dimensional groundwater flow equation using the finite difference method. Similar to previous versions of MODFLOW such as MODFLOW-2005 (Harbaugh, 2005), MODFLOW-NWT can represent different components of the groundwater flow system, including recharge, groundwater withdrawal from wells, and groundwater/surface-water interaction. However, MODFLOW-NWT uses a different numerical technique—the Newton-Raphson method—for solving the groundwater flow equation and requires a different intercell conductance package—the Upstream Weighting (UPW) package—to guarantee mass balance. Using MODFLOW-NWT improves model convergence for groundwater problems in unconfined settings with substantial groundwater/surface-water interaction.

Model Framework

Historical groundwater-level data indicate that under natural conditions, the groundwater system in the Matanuska-Susitna Valley is in dynamic steady state. Interannual groundwater-level fluctuations of several feet were observed in response to climatic forcing; however, average groundwater levels in monitor wells remained constant at decadal time scales. Time-series groundwater-level data collected during this study show similar patterns of temporal variability to historical groundwater-level data during periods of similar precipitation deficits. Furthermore, field investigations of groundwater/surface-water interaction in the study area suggest that rates of groundwater discharge to surface water are at least one order of magnitude higher than groundwater withdrawals from wells. An analysis of groundwater age (Glass, 2001) shows that groundwater residence time is less than 100 years in most of the core area, and the hydraulic conductivity of primary water-supply aquifers is relatively high – generally 100 ft/d or higher. It is likely that changes in groundwater flow patterns induced by hydrologic stresses propagate quickly through the groundwater system, such that increased outflows to wells are offset by reduced outflows to other system boundaries. For these reasons, a steady-state simulation was used to simulate groundwater flow in the Matanuska-Susitna Valley, solving the groundwater flow equation subject to the condition that inflows are equal to outflows and there is no change in storage within the system.

Spatial Discretization

A model grid of 125 rows, 150 columns, and 3 layers was used to represent the groundwater system of the Matanuska-Susitna Valley. Each model cell has dimensions of 2,000 by 2,000 ft in the horizontal plane and variable height in the vertical direction. Cells outside the model area were designated as inactive in MODFLOW-NWT. The model layers were defined in such a way to represent the hydrogeology of the study area. Cells in model layer 1 represent three hydrogeologic units: Holocene Outwash and Alluvium, Fine Sediments, and Naptowne Moraine. Cells in model layer 1 intersecting the Naptowne Moraine and the less permeable zone of the Fine Sediments hydrogeologic unit, east of the Little Susitna River, were similarly grouped within a single zone (HK1-1) and assigned uniform hydraulic properties (fig. 27A). Cells in model layer 1 intersecting the Holocene Outwash and Alluvium and the more permeable deposits of the Fine Sediments hydrogeologic unit, west of the Little Susitna River, were grouped within a second zone (HK1-2) with uniform hydraulic properties. Cells in model layer 2 represent two hydrogeologic units: Lower Permeable Sediments and Tertiary Sedimentary Rocks. Cells in model layer 2 intersecting the Lower Permeable Sediments hydrogeologic unit were grouped within a single zone (HK2-1) and assigned uniform hydraulic properties; similarly, those cells intersecting the Tertiary Sedimentary Rocks hydrogeologic unit were grouped within a second zone (HK2-2) and assigned uniform hydraulic properties (fig. 27B). Cells in model layer 3 represent the Tertiary Sedimentary Rocks hydrogeologic unit and were assigned identical hydraulic properties as cells in the second zone of model layer 2. Each hydrogeologic unit is recognized to be internally heterogeneous; this heterogeneity was simulated by using effective hydraulic properties within each zone.

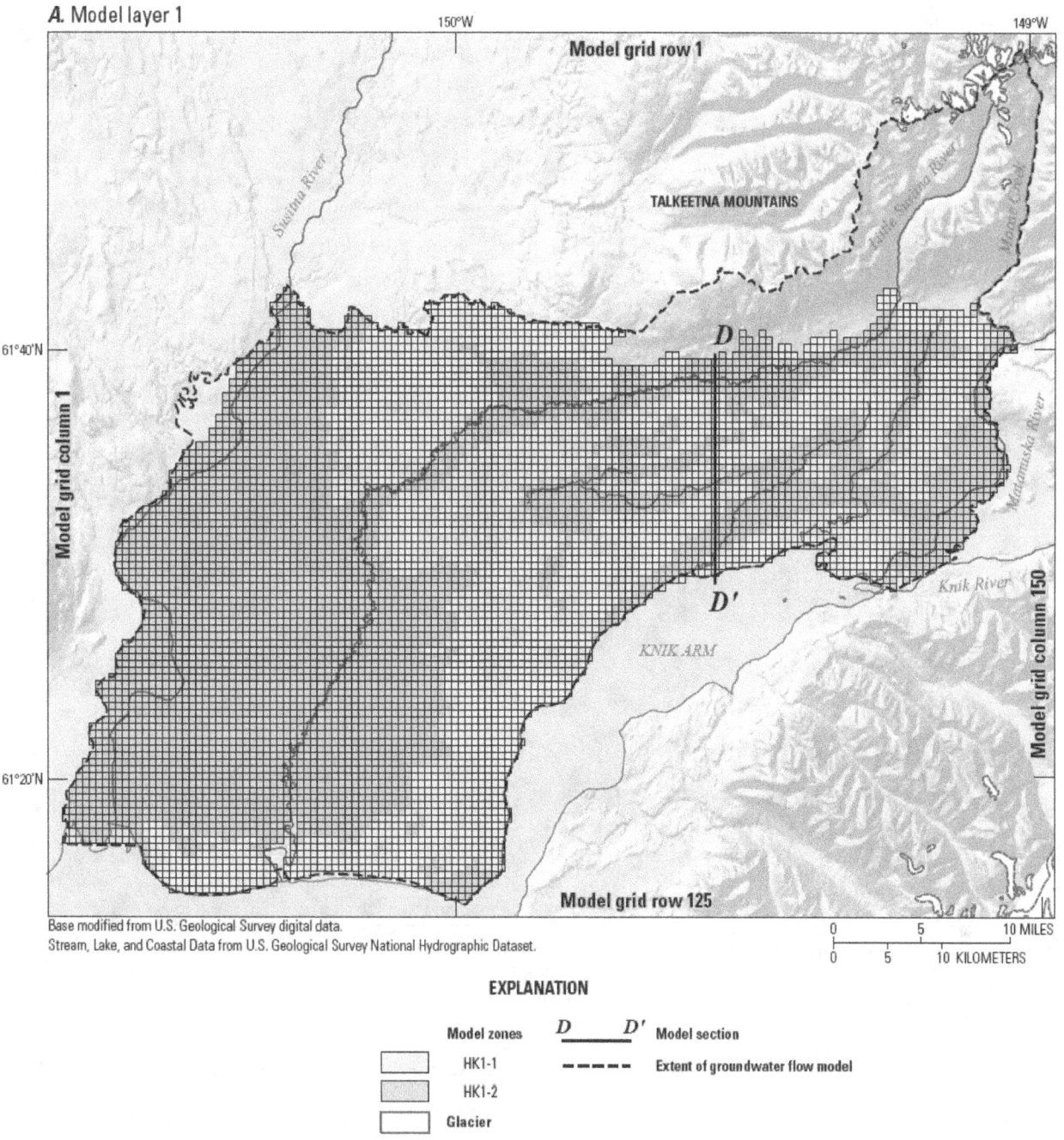

Figure 27. Hydraulic conductivity zones in model layers 1-2, Matanuska-Susitna Valley, Alaska. (*A*) Hydraulic conductivity zones in model layer 1, HK1-1 is horizontal hydraulic conductivity in zone 1, HK1-2 is horizontal hydraulic conductivity in zone 2; (*B*) Hydraulic conductivity zones in model layer 2, HK2-1 is horizontal hydraulic conductivity in zone 1, HK2-2 is horizontal hydraulic conductivity in zone 2.

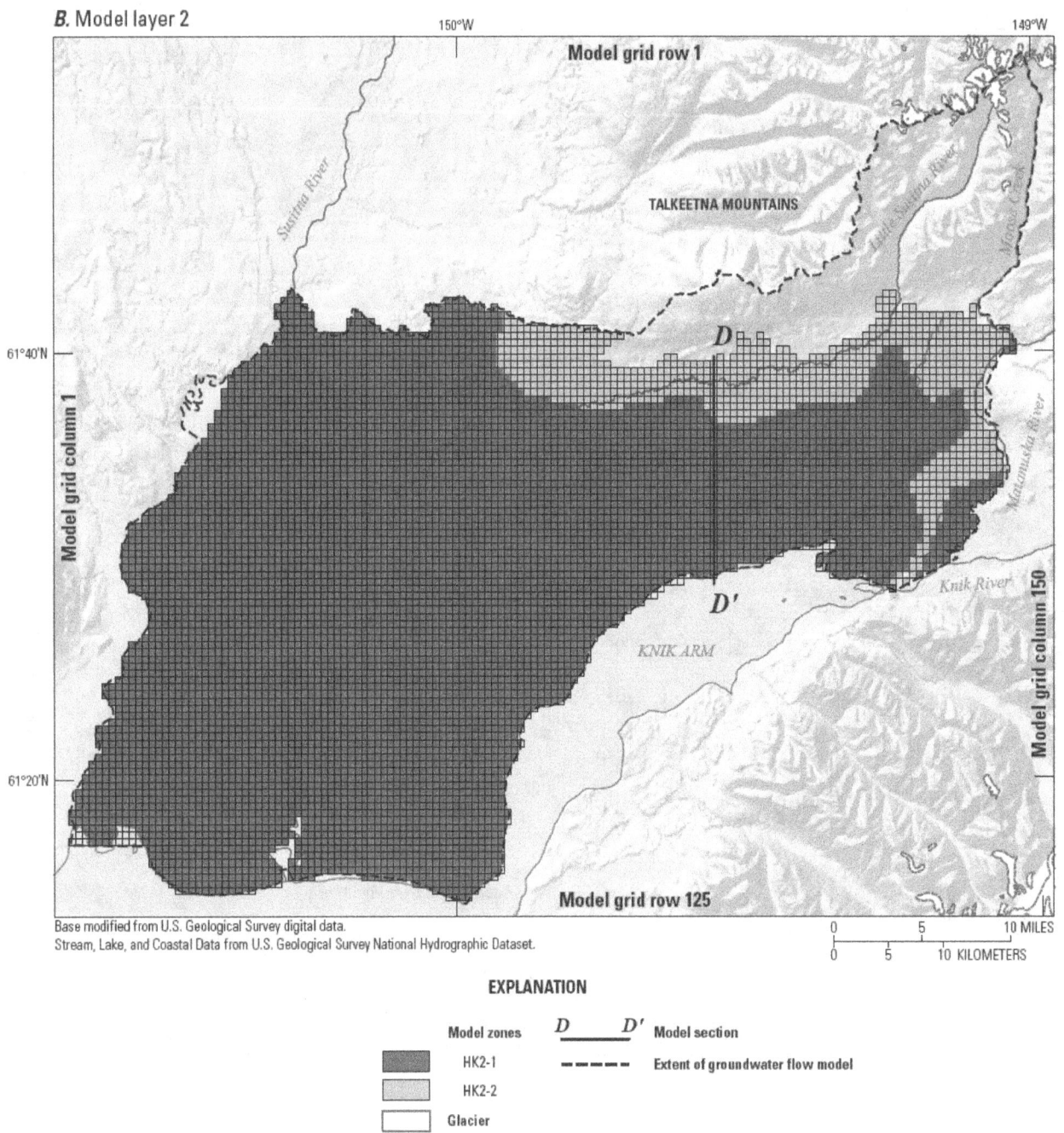

B. Model layer 2

Figure 27. Continued.

In general, unconsolidated geologic strata and sedimentary rocks are anisotropic, with the horizontal hydraulic conductivity much greater than the vertical hydraulic conductivity (Freeze and Cherry, 1979). The Tertiary Sedimentary Rocks unit consists of interbedded layers of rock types with different hydraulic properties. Water-bearing fractures and seeps in this unit are commonly encountered within sandstone or coal beds but not within siltstone or shale beds. Therefore, it is reasonable to treat this hydrogeologic unit as anisotropic. The ratio of horizontal to vertical hydraulic conductivity $(K_h : K_v)$ for the Tertiary Sedimentary Rocks hydrogeologic unit was set equal to 10. Similarly, the unconsolidated hydrogeologic units, including glacial, alluvial, lacustrine, and estuarine sediments, were assumed to be anisotropic. A typical value for $K_h : K_v$, is 10:1. In the absence of detailed information for the hydrogeologic units represented in the groundwater flow model, the ratio $K_h : K_v$

was set as 10:1 for all active cells in the groundwater model grid. It was further assumed that the directions of the model grid are collinear with the principal directions of horizontal and vertical hydraulic conductivity.

MODFLOW-NWT requires that model layers be specified as either confined or convertible; for convertible layers, formulation of cell conductance terms in the groundwater flow equation changes when hydraulic head decreases below the top of a cell. For this model, model layer 1 was specified as a convertible layer, and model layers 2 and 3 were specified as confined layers. The bottom elevations for cells in each model layer were assigned according to the results of the hydrogeologic framework model. Section D-D' (fig. 28) shows how the bottom elevations of cells in each model layer were assigned in order to accurately represent the hydrogeologic framework model for the Matanuska-Susitna Valley groundwater system.

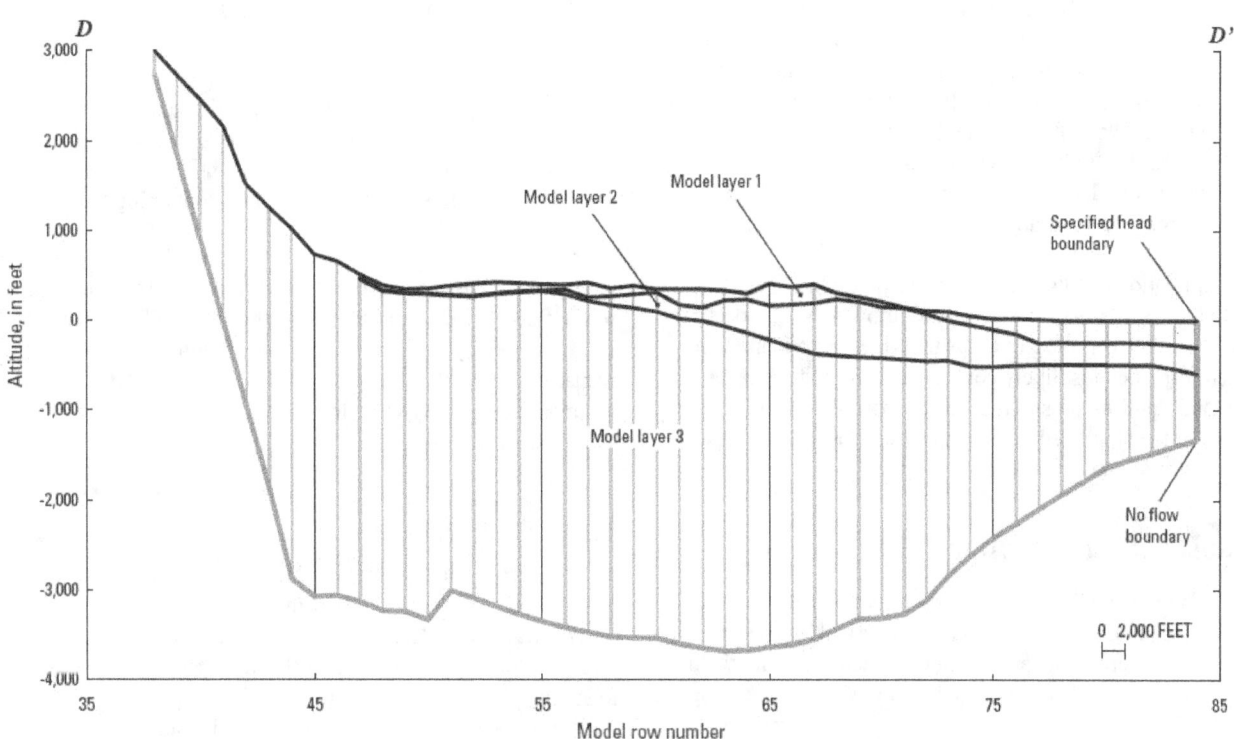

Figure 28.   Model section D-D' showing variability of thickness in model layers and the boundary conditions assigned to represent the salt–water/freshwater interface at Knik Arm, Matanuska-Susitna Valley, Alaska.

Model Boundaries and Implementation of MODFLOW Packages

The extent of the simulated groundwater flow system was bounded by physically meaningful hydrologic features. In addition to boundaries at the edges of the simulated area, stress packages in MODFLOW-NWT were used to represent other important hydrologic features in the Matanuska-Susitna Valley groundwater system, including surface-water features, pumping wells, and in-place recharge. The extent of active cells in each model layer and locations of various model boundaries are shown in figure 29.

Numerous surface-water features—including lakes, wetlands, ponds, and small streams—are located within the modeled area. It is likely that some or all these features are connected to the groundwater system. However, for many of these surface-water features in remote areas, little or no hydrologic information was available. Furthermore, the area of primary interest for the objectives of this study was the populated area of the Matanuska-Susitna Valley, including the cities of Palmer, Wasilla, and Houston and the MSB core area. As a result, many of the surface-water features within the modeled area were not explicitly considered as boundaries in the groundwater flow model. Notably, this includes lake-wetland complexes west of Big Lake and throughout the Susitna Lowlands. These surface-water features are distant from the populated area, so their omission from the groundwater flow model should have negligible influence on the simulation results in the populated area. In the western part of the modeled area, groundwater outflow may only occur to the Susitna River, the Little Susitna River, or to Knik Arm. Therefore, the most likely consequence of omitting lakes, wetlands, and streams in the western part of the modeled area was that the simulated outflow to these features was larger than the true outflow.

Extent of Simulated Area

The Susitna River forms the western boundary of the simulated area, and the Matanuska River and Moose Creek form the eastern boundary of the simulated area. The northern boundary of the simulated area is a watershed divide along the ridge of the Talkeetna Mountains, and the southern boundary is Knik Arm. The Matanuska River and the Susitna River were simulated as head-dependent flow boundary conditions using the River Package. The watershed divide along the Talkeetna Mountains was simulated as a no-flow boundary. In coastal environments, groundwater discharge to the ocean occurs primarily along seepage faces and as shallow submarine groundwater discharge. Along the coast, less-dense freshwater flows over more-dense salt water before discharging to the ocean. The zone of mixing between freshwater and salt water is commonly represented as a sharp interface. To simulate flow near the freshwater/salt-water interface, a specified-head boundary was assigned to cells adjacent to Knik Arm in model layer 1. A no-flow boundary was assigned to cells adjacent to Knik Arm in model layers 2 and 3. This representation of the freshwater/salt-water interface is illustrated in section D-D' (fig. 28).

Recharge Package

The Recharge Package was used to simulate in-place recharge from precipitation and septic effluent as a specified-flow boundary. In the Recharge Package, in-place recharge is computed as

$$QR = I \times DELR \times DELC \qquad (3)$$

where
  QR is the recharge flow applied to a model cell,
  I is the recharge flux for the model cell,
  $DELR$ is the width of the row containing the cell, and
  $DELC$ is the width of the column containing the cell.

Values of mean annual recharge for the period 2002–10—based on the DPM and estimates of septic and irrigation return flows—were used to specify the recharge rate to each cell in model layer 1. The spatial distribution of groundwater recharge is variable; the highest recharge rates are in the mountainous headwater valleys of the Little Susitna River and Moose Creek.

Well Package

The Well Package was used to simulate groundwater withdrawals for supply wells with data on groundwater withdrawal rates. Mean annual groundwater withdrawal rates for 2004–10 based on data shown in table 5 were used to specify outflow rates applied to each cell containing a pumping well. Also, groundwater withdrawals from domestic wells were estimated as described in the section, "Groundwater Withdrawals." The outflow rates associated with groundwater withdrawals for each parcel were then assigned to the model cell enclosing the parcel centroid.

A. Model layer 1

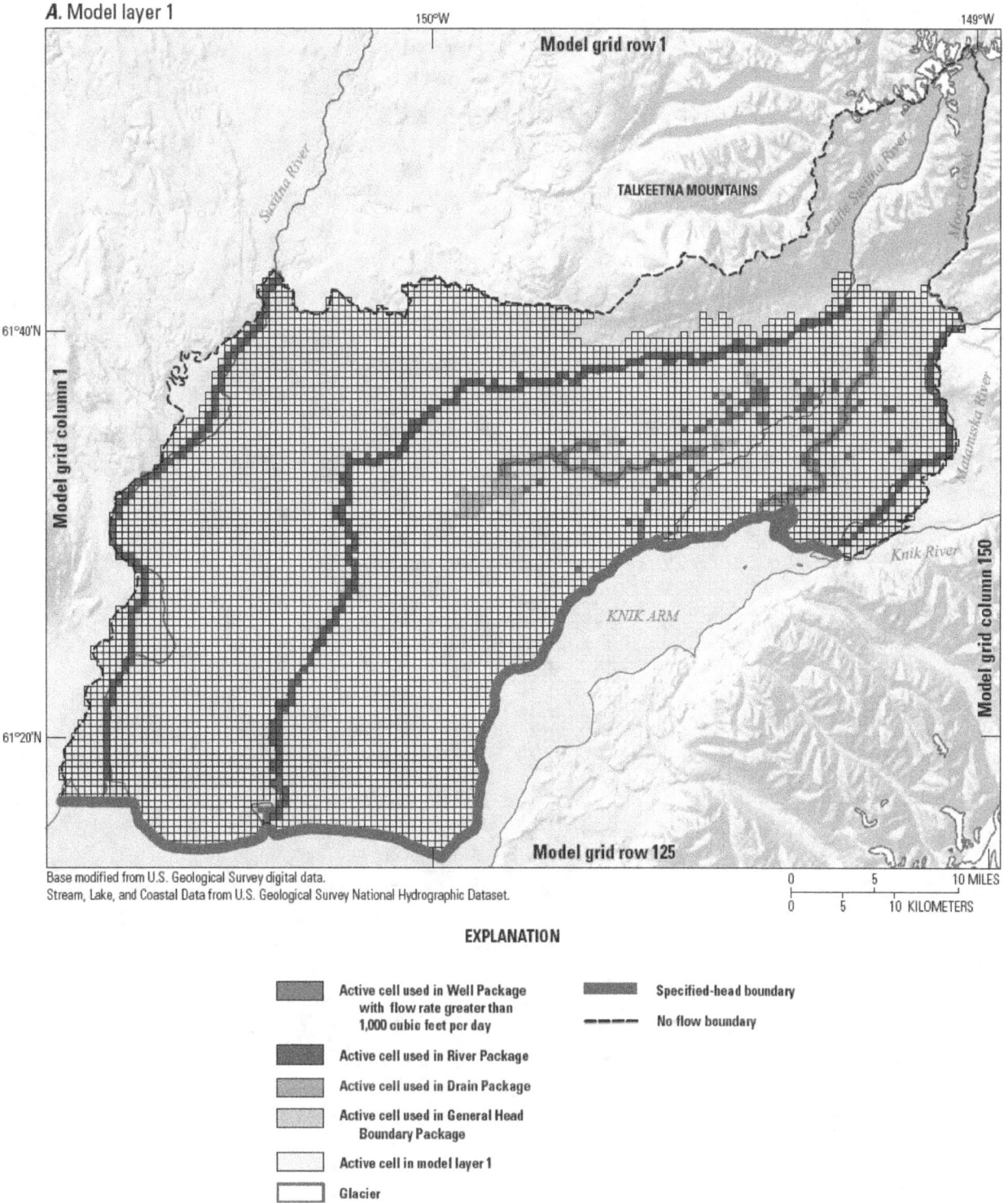

Figure 29.   Active cells and model boundaries for groundwater flow model, Matanuska-Susitna Valley, Alaska. (*A*) Active cells in model layer 1; (*B*) Active cells in model layer 2; (*C*) Active cells in model layer 3.

B. Model layer 2

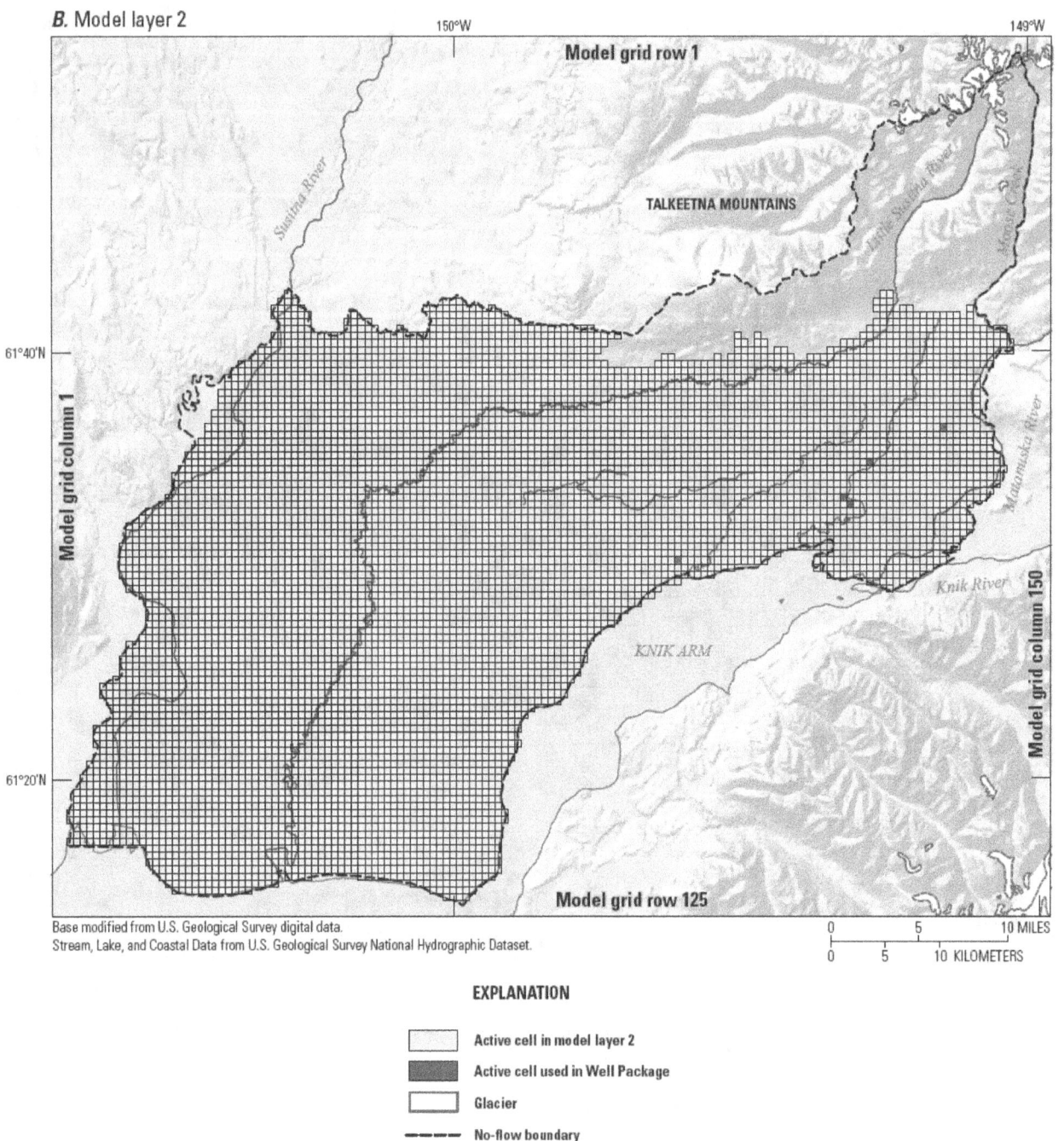

Base modified from U.S. Geological Survey digital data.
Stream, Lake, and Coastal Data from U.S. Geological Survey National Hydrographic Dataset.

0   5   10 MILES

0   5   10 KILOMETERS

EXPLANATION

Active cell in model layer 2

Active cell used in Well Package

Glacier

------- No-flow boundary

Figure 29.—Continued

C. Model layer 3

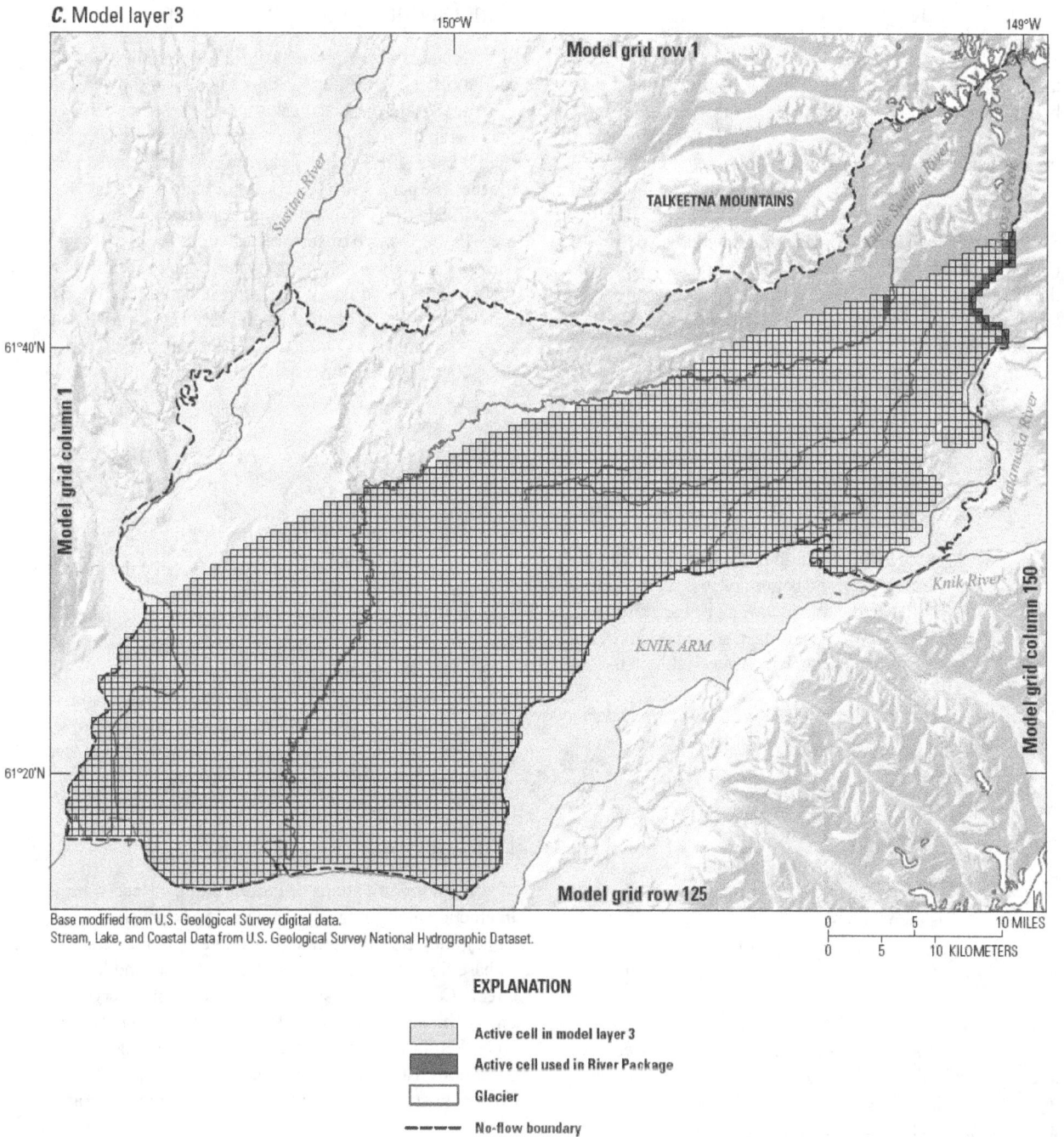

Base modified from U.S. Geological Survey digital data.
Stream, Lake, and Coastal Data from U.S. Geological Survey National Hydrographic Dataset.

0   5   10 MILES
0   5   10 KILOMETERS

EXPLANATION

Active cell in model layer 3

Active cell used in River Package

Glacier

------ No-flow boundary

Figure 29.—Continued

River Package

Moose Creek, Matanuska River, Little Susitna River, and Susitna River were represented as head-dependent flow boundaries using the River Package. The River Package boundaries were assigned for the model layer cell intersecting the river. Most of the river segments represented using the River Package intersect model layer 1 of the groundwater flow model. However, several reaches of Moose Creek and Little Susitna River intersect the Tertiary Sedimentary Rocks unit (fig. 29C); those reaches were assigned to model layer 3. In the River Package, the volumetric flow rate across the riverbed between the river and the underlying groundwater model cell is computed as

$$Q_{rb} = C_{rb}(h_r - h_a) \tag{4}$$

where

Q_{rb} is the flow rate across the riverbed,
C_{rb} is the hydraulic conductance of the riverbed,
h_r is the river stage, and
h_a is the hydraulic head in the cell underlying the riverbed if the bottom of the riverbed is below the water table in the cell or the altitude of the bottom of the river bed if the bottom of the riverbed is above the water table in the cell

The hydraulic conductance of the riverbed is computed as

$$C_{rb} = \frac{K_v wL}{m} \tag{5}$$

where

K_v is the vertical hydraulic conductivity of the riverbed sediment,
w is the width of the river reach,
L is the length of the river reach, and
m is the thickness of the riverbed sediments.

In the River Package, a river reach refers to the section of a river within a model cell. The river stage in each river was approximated using USGS topographic maps and checked against gage elevations measured at USGS streamgages along the river. The widths of river reaches were determined from aerial photos at control points along the river, and the riverbed thickness was set to vary linearly from 10 to 20 ft with longitudinal distance along the river. River segments were extracted from the National Hydrographic Dataset (U.S. Geological Survey, 2009), and the computer program RIVGRID (Leake and Claar, 1999) was used to generate input files for the River Package using data sources described above. During model calibration, the riverbed vertical hydraulic conductivity was varied to improve the fit between simulated and observed hydraulic heads.

General Head Boundary Package

Examination of groundwater-level variability in nine reference lakes over the 3-year study period showed that several lakes in the Matanuska-Susitna Valley are important features in the regional groundwater flow system. Specifically, Lucile Lake and Blodgett Lake are groundwater fed during the ice-free period and likely during the ice-affected period as well. Big Lake drains the entire area west of the MSB core area. In the absence of surface-water outflows, drainage from Memory Lake is an important water input to the groundwater system. These four lakes were represented as head-dependent flow boundaries using the General Head Boundary Package. In the General Head Boundary Package, flow between the General Head Boundary and underlying aquifer cell is computed as

$$Q_{GHB} = CB(h_{GHB} - h_a) \tag{6}$$

where

Q_{GHB} is the flow rate across the boundary,
CB is the hydraulic conductance of the boundary,
h_{GHB} is the hydraulic head assigned to the boundary, and
h_a is the hydraulic head in the cell connected to the boundary.

The flow between General Head Boundaries and connected aquifer cells is linear and continuous; also, water is allowed to flow either into or out of connected aquifer cells without limit.

Drain Package

Field investigations of groundwater/surface-water interaction along small streams in the Matanuska-Susitna Valley showed groundwater discharge is an important source of base flow in Wasilla Creek, Lucile Creek, and Meadow Creek. Consistently losing reaches were not observed in any of these streams during the study period. These streams were therefore represented as head-dependent flow boundary conditions using the Drain Package. In the Drain Package, flow from an aquifer cell into a connected drain is computed as

$$Q_D = \begin{cases} CD(h_a - h_D), & h_a > h_D \\ 0, & h_a \le h_D \end{cases} \tag{7}$$

where

Q_D is the flow rate from the aquifer to the drain,
CD is the hydraulic conductance of the drain,
h_D is the drain elevation, and
h_a is the hydraulic head in the cell containing the drain.

Flow from aquifer cells in the drains is piecewise linear and nonzero only when the head in a model cell is greater than the drain elevation. The Drain Package allows for groundwater outflow to drains but not groundwater inflow from drains. This is therefore an appropriate stress package to use for springs or constantly gaining streams, such as those described above. The values of drain conductance for each stream were adjusted during the model calibration process to match the groundwater outflows determined from field measurements.

Model Calibration

Model calibration is the process of adjusting of model parameters—including aquifer hydraulic properties for different hydrogeologic units and hydraulic conductance of various model boundaries—to minimize the difference between observed and simulated quantities in the groundwater system. Model calibration was performed using UCODE-2005 (Poeter and others, 2005). UCODE-2005 uses nonlinear regression techniques to estimate parameter values that minimize mismatch between observed and simulated quantities. Specific details of the nonlinear regression techniques are described by Poeter and others (2005) and Hill and Tiedeman (2007). Model calibration is assessed by examining agreement between those observed and simulated quantities.

Calibration Data

In this study, water levels and flows between groundwater and surface water were used as observations in the model-calibration procedure. The water levels used in model calibration included 135 groundwater-level measurements and 34 lake-stage measurements made during the synoptic water-level survey in summer 2009. Big Lake, Lucile Lake, Blodgett Lake, and Memory Lake were represented as General Head Boundaries in the groundwater flow model. Therefore, water-level measurements from those lakes were not included as head observations during the model-calibration procedure but, rather, were input as boundary conditions to the model. Flows between groundwater and surface water were defined using reach-scale gain or loss estimates from seepage runs performed on Wasilla Creek, Lucile Creek, and Meadow Creek during 2009–11. Specifically, six flow observations were available for Wasilla Creek, seven flow observations were available for Lucile Creek, and five flow observations were available for Little Meadow and Meadow Creek.

Seepage runs were performed on multiple occasions for all streams; to be applicable to the steady state model, these data were averaged to represent average conditions.

UCODE-2005 requires that the user specify weighting values for each observation used in the regression; typically, those weighting values are calculated from the estimated error of each observation. Error in groundwater-level measurements is calculated as the sum of error in the land-surface elevation or top of well casing and error in depth to water. For this study, error in the elevation of the top of the well casing was +0.1 ft, and error in depth to water measurement was +0.01 ft, resulting in an estimated error of +0.1 ft for each groundwater-level measurement. Error in lake-stage measurements depends only on error in the lake-stage elevation and is estimated as +0.1 ft. Treating each of these values as 95-percent confidence intervals on normally distributed random variables (Hill, 1998), their respective standard deviations are both 0.05 ft.

Errors in flow observations were calculated similarly to errors in water-level observations. Most streamflow measurements made during this study were rated with measurement errors less than 10 percent of the reported value. As in the case of water-level measurements, the streamflow measurement error was treated as a 95-percent confidence interval, allowing for calculation of standard deviations and variances for each streamflow measurement. Streamflow gains and losses are calculated by subtracting one streamflow measurement from another streamflow measurement. The uncertainty in streamflow gains and losses is calculated by summing the variances of two or more streamflow measurements used to estimate gains and losses (Hill, 1998).

Estimated Parameters

The groundwater flow model includes 13 different parameters corresponding to horizontal hydraulic conductivity of different zones and values of boundary conductance, including riverbed conductance, drain conductance, and general head-boundary conductance (table 12). All the model parameters were estimated using UCODE-2005.

For segments of the Little Susitna River in contact with the Tertiary Sedimentary Rocks hydrogeologic unit, the river cuts directly into bedrock. Therefore, the riverbed vertical hydraulic conductivity was set equal to the vertical hydraulic conductivity of the Tertiary Sedimentary Rocks hydrogeologic unit. For all other river segments, the value of the riverbed vertical hydraulic conductivity was estimated in the regression. General Head Boundary conductance values were estimated independently for each of the four lakes represented in the groundwater flow model.

Table 12.   Estimated values for horizontal hydraulic conductivity, boundary conductance, and river vertical hydraulic conductivity, Matanuska-Susitna Valley, Alaska.

[**Abbreviations:** ft/d, foot per day; ft²/d, food squared per day]

Parameter	Corresponding geologic unit or material	Parameter type	Value
HK1-1	Naptowne Moraine and Fine-Grained Sediments	Horizonal hydraulic conductivity (ft/d)	16.9
HK1-2	Holocene Outwash and Alluvium	Horizonal hydraulic conductivity (ft/d)	19.3
HK2-1	Lower Permeable Sediments	Horizonal hydraulic conductivity (ft/d)	0.35
HK2-2	Tertiary Sedimentary Rocks	Horizonal hydraulic conductivity (ft/d)	0.02
HK3	Tertiary Sedimentary Rocks	Horizonal hydraulic conductivity (ft/d)	0.02
Cond-WC	Wasilla Creek streambed sediments	Drain conductance (ft²/d)	1.6E+07
Cond-MC	Meadow Creek streambed sediments	Drain conductance (ft²/d)	1.57E+09
Cond-LC[1]	Lucile Creek streambed sediments	Drain conductance (ft²/d)	3.29E+06
Cond-RIV[1]	Matanuska River, Moose Creek, Little Susitna River, Susitna River riverbed sediments	River vertical hydraulic conductivity (ft/d)	1.0
Cond-BgL	Big Lake lakebed sediments	General head boundary conductance	5.0E+05
Cond-LL	Lucile Lake lakebed sediments	General head boundary conductance	5.0E+05
Cond-BlL	Blodgett Lake lakebed sediments	General head boundary conductance	5.0E+05
Cond-ML	Memory Lake lakebed sediments	General head boundary conductance	5.0E+05

[1]Conductance estimated by varying vertical hydraulic conductivity with riverbed area calculated directly and fixed riverbed thickness.

The horizontal hydraulic conductivity values estimated for zones 1 and 2 of model layer 1, corresponding to Naptowne Moraine and Holocene Outwash and Alluvium hydrogeologic units, were quite similar (table 12). The estimated value for the Lower Permeable Sediments hydrogeologic unit was less than the values for the overlying units but greater than the fixed value for the Tertiary Sedimentary Rocks hydrogeologic unit. It is important to note that these are all effective values applied over a relatively large area. The estimated value for drain conductance corresponding to Wasilla Creek was several orders of magnitude less than the values estimated for Meadow Creek. Horizontal hydraulic gradients are generally steeper in the area surrounding Wasilla Creek than in the area surrounding Lucile and Meadow Creek. The lower drain conductance values for Wasilla Creek may then serve to reduce overestimation of groundwater outflow to Wasilla Creek.

Sensitivity Analysis

Reliable parameter estimation requires that observation data contain sufficient information to uniquely identify each estimated parameter. Insensitivity to model parameters and parameter correlation are two common problems that may arise during the nonlinear regression procedure. UCODE-2005 calculates a number of statistics that may be used to evaluate the reliability of parameter estimates obtained during the nonlinear regression procedure, two of which pertain specifically to the problems of sensitivity to model parameters and parameter correlation.

Composite Scaled Sensitivities

If model-simulated values are insensitive to parameter changes, observations corresponding to those simulated values are uninformative for the purposes of parameter estimation (Hill and Tiedeman, 2007). Insensitivity to model parameters may prevent the nonlinear regression procedure from converging, lead to excessively large confidence intervals on the estimated parameters, or both. The composite scaled sensitivity (CSS) is a statistic commonly used to evaluate the sensitivity of model-simulated values to changes in the parameters. In UCODE-2005, the CSS is computed as

$$css_j = \sum_{i=1}^{ND} \left[\frac{\left(dss_{ij}\right)^2 |_b}{ND} \right]^{1/2} \tag{8}$$

where

css_j is the composite scaled sensitivity of the *jth* parameter,

ND is the number of observations used in the regression procedure,

dss_{ij} is the dimensionless scaled sensitivity of the *ith* observation to the *jth* parameter, and

b is the vector of parameters specifying the location in the parameter space at which the dimensionless scaled sensitivity is evaluated.

and the dimensionless scaled sensitivities (DSS) are computed as

$$dss_{ij} = \left[\frac{\partial y'_i}{\partial b_j} \Big|_b \, b_j \right] \left| \ln\left(b_j\right) \right| \omega_{ii}^{\frac{1}{2}} \tag{9}$$

where

$\dfrac{\partial y'_i}{\partial b_j} \Big|_b$ is the sensitivity of the *ith* simulated value with respect to the *jth* parameter, evaluated at the set of parameter values in **b,**

b_j is the *jth* estimated parameter, and

ω_{ii} is the weight assigned to the *ith* observation

Hill and Tiedeman (2007) suggest that a threshold value of 1.0 be used to identify parameters for which available observations are not sufficiently informative to reliably estimate the parameter value. The CSS values computed for each parameter at the final set of parameter values for which the regression procedure converged are shown in table 13. Using the suggested threshold value of 1.0 indicates that the observations are sufficiently informative to estimate the values for the parameters HK1-1, HK1-2, HK2-1, HK2-2, HK3, and Cond-MC. For the parameters Cond-WC, Cond-LC, Cond-RIV, Cond-BgL, Cond-LL, CondBlL, and Cond-ML, the CSS values are less than 1.0, indicating that the observations are not sufficiently informative to estimate their respective values. The latter set of parameters comprises boundary conductance parameters. In UCODE-2005, parameters with vanishingly small CSS values are omitted from the regression procedure. In this case, the insensitive boundary conductance parameters listed above were all omitted.

In general, flow rates in a groundwater flow model are sensitive to boundary conductance parameters only when the parameter values are small (Tucci, 1982). The calibrated values of boundary conductance parameters obtained in the regression typically fell outside of the sensitive range and therefore, it may be unreasonable to expect reliable estimation of boundary conductance parameters by regression procedures. However, the sensitivity statistics presented here suggest that the boundary conductance parameters have less of an impact on simulated heads and flows than the other parameters and may therefore not be as important for the purposes of the model.

Parameter Correlation Coefficients

Correlated parameters are those for which a change in the value of one parameter follows changes in the value of another parameter in a predictable way. In the case of extreme parameter correlation, it may be possible to estimate the ratio of the correlated parameters but not the actual

Table 13. Composite scaled sensitivity values for each model parameter estimated during the nonlinear regression procedure, computed in UCODE-2005

Parameter	Composite scaled sensitivity value
HK1-1	282.7
HK1-2	325.8
HK2-1	44.26
HK2-2	5.474
HK3	5.474
Cond-WC	0.354
Cond-MC	18.06
Cond-LC	0.548
Cond-RIV	0.0001
Cond-BgL	0.0
Cond-LL	0.0
Cond-BlL	0.0
Cond-ML	0.0

parameter values themselves. A common way to evaluate the magnitude of parameter correlation is through the calculation of parameter correlation coefficients (PCC). In UCODE-2005, the PCCs are computed as

$$pcc(j,k) = \frac{Cov(j,k)}{Var(j)^{1/2}\,Var(k)^{1/2}} \tag{10}$$

where

$pcc(j,k)$ is the parameter correlation coefficient between the *jth* and *kth* parameters,

$Cov(j,k)$ is the covariance between the *jth* and *kth* parameters, and

$Var(j)$ is the variance of the *jth* parameter.

PCC values range from -1.00 to +1.00, with more extreme parameter correlation as the absolute value of the PCC approaches 1.00. In general, values for parameter pairs with PCC values between -0.95 and +0.95 can be uniquely estimated (Hill and Tiedeman, 2007). The PCC values computed during the regression procedure for this groundwater flow model all fall between -0.85 and +0.85, indicating that it was possible to uniquely estimate each of the model parameters during the regression procedure. The PCC values for each parameter pair are displayed in table 14. The PCC values are all zero for pairs of boundary conductance parameters identified as insensitive from analysis of CSS statistics, reflecting the omission of these insensitive parameters from the regression procedure.

Table 14.   Parameter correlation coefficients for pairings of each parameter estimated during the nonlinear regression procedure, computed using UCODE-2005.

	HK1-1	HK1-2	HK2-1	HK3	Cond-WC	Cond-LC	Cond-MC	Cond-RIV	Cond-BgL	Cond-LL	Cond-BIL	Cond-ML
HK1-1	1.000	-0.402	0.019	-0.360	0.005	-0.011	0.000	-0.001	0.000	0.000	0.000	0.000
HK1-2	-0.402	1.000	-0.079	-0.229	-0.073	-0.012	0.001	0.008	0.000	0.000	0.000	0.000
HK2-1	0.019	-0.079	1.000	-0.434	-0.014	0.006	0.000	0.002	0.000	0.000	0.000	0.000
HK3	-0.360	-0.229	-0.434	1.000	0.039	-0.021	0.000	-0.006	0.000	0.000	0.000	0.000
Cond-WC	0.005	-0.073	-0.014	0.039	1.000	-0.024	-0.009	-0.117	0.000	0.000	0.000	0.000
Cond-LC	-0.011	-0.012	0.006	-0.021	-0.024	1.000	0.004	0.056	0.000	0.000	0.000	0.000
Cond-MC	0.000	0.001	0.000	0.000	-0.009	0.004	1.000	0.076	0.000	0.000	0.000	0.000
Cond-RIV	-0.001	0.008	0.002	-0.006	-0.117	0.056	0.076	1.000	0.000	0.000	0.000	0.000
Cond-BgL	0.000	0.000	0.000	0.000	0.000	0.000	0.000	0.000	1.000	0.000	0.000	0.000
Cond-LL	0.000	0.000	0.000	0.000	0.000	0.000	0.000	0.000	0.000	1.000	0.000	0.000
Cond-BIL	0.000	0.000	0.000	0.000	0.000	0.000	0.000	0.000	0.000	0.000	1.000	0.000
Cond-ML	0.000	0.000	0.000	0.000	0.000	0.000	0.000	0.000	0.000	0.000	0.000	1.000

Assessment of Calibration

The distribution of hydraulic heads computed in layer 1 of the groundwater flow model is similar to the water-table contour maps synthesized from previous studies and synoptic water-level measurements performed during this study (fig. 30A). In particular, the model results qualitatively match water-table mounding near Memory Lake and southwest of Wasilla. The orientation of water-level contours indicates that groundwater discharges into the Susitna River, Little Susitna River, and Matanuska River along the entire length of each respective river. The simulated groundwater budget provides a reasonable match to independently estimated groundwater budget components. In the section "Simulated Heads," the model is assessed by comparing simulated and observed hydraulic heads and by examining the spatial distribution of residuals as well as by comparing simulated and observed flows between groundwater and surface water.

Simulated Heads

Hydraulic-head residuals are defined as the difference between observed and simulated hydraulic heads; a positive value indicates that the model-simulated head is too low, and a negative value means that the model-simulated head is too high. The mean and standard deviation of the hydraulic-head residuals calculated from the calibrated groundwater flow model are -2.6 and 23.8 ft, respectively. Error in model-simulated heads may result from incorrect hydrogeologic structure, boundary conditions, or hydraulic properties of the aquifer. Comparison of head residuals with observed heads shows a very slight positive relation between residuals and observed heads, meaning that the groundwater flow model may have a general tendency toward higher estimates of smaller heads and lower estimates of larger heads (fig. 31).

Contrary to this general trend, it should also be noted that the largest observed heads are associated with a negative value on the residual.

Plotting the spatial distribution of residuals provides another valuable tool for assessing the groundwater flow model. Patterns in the spatial distribution of the residuals may indicate model structural deficiencies. Across most of the modeled area, the positive- and negative-signed residuals are equally distributed (fig. 32). However, there is a northeast-southwest trending band of large positive-signed residuals between the Little Susitna River and the city of Wasilla. This band corresponds to topographic and water-table ridges. The water-table ridge likely forms a groundwater divide between the Little Susitna River subwatershed to the north and Cottonwood and Meadow Creek subwatersheds to the south. Observed patterns in the spatial distribution of the residuals indicate that the groundwater flow model is unable to accurately simulate this feature. This discrepancy between observed and simulated heads could result from insufficiently detailed representation of groundwater recharge. Groundwater recharge simulated with the DPM is spatially uniform in the HUC-12s immediately south of the Little Susitna River; it is possible that local-scale variations may be present in land use and land cover, or in soil characteristics, which are not adequately represented by the DPM. Alternately, the hydraulic head discrepancy could result from lateral variation in the hydraulic conductivity of the Naptowne Moraine that is not accurately captured by the zonation scheme in model layer 1. A similar water-table ridge is further south, trending in the same direction, between the city of Wasilla and Knik Arm. Spatial patterns in the distribution of residuals are not observed in this part of the model domain, indicating that the groundwater flow model accurately simulated this feature.

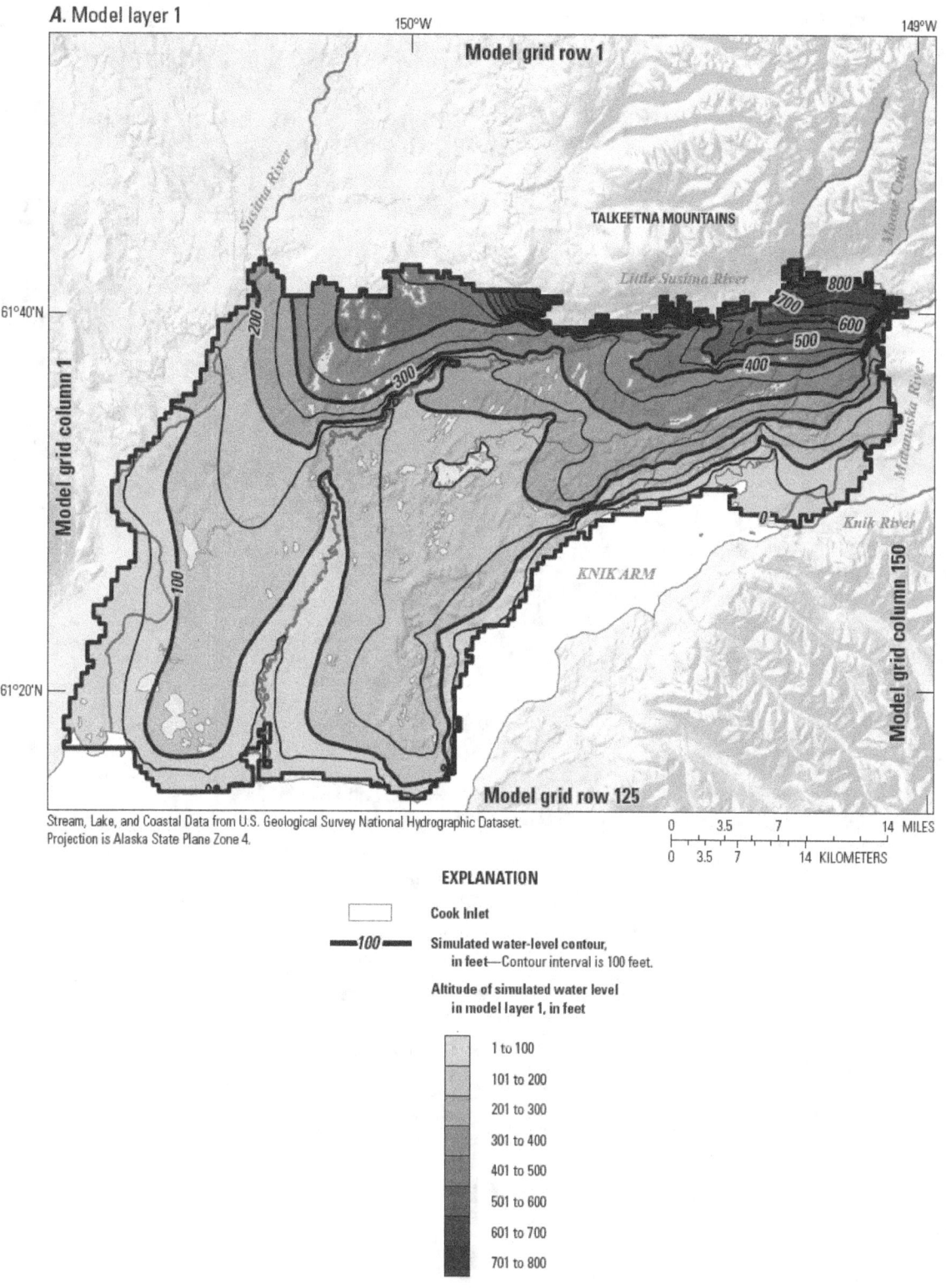

A. Model layer 1

Figure 30. Simulated water levels in the Matanuska-Susitna Valley, Alaska. (*A*) Simulated water levels in model layer 1; (*B*) Simulated water levels in model layer 2; (*C*) Simulated water levels in model layer 3.

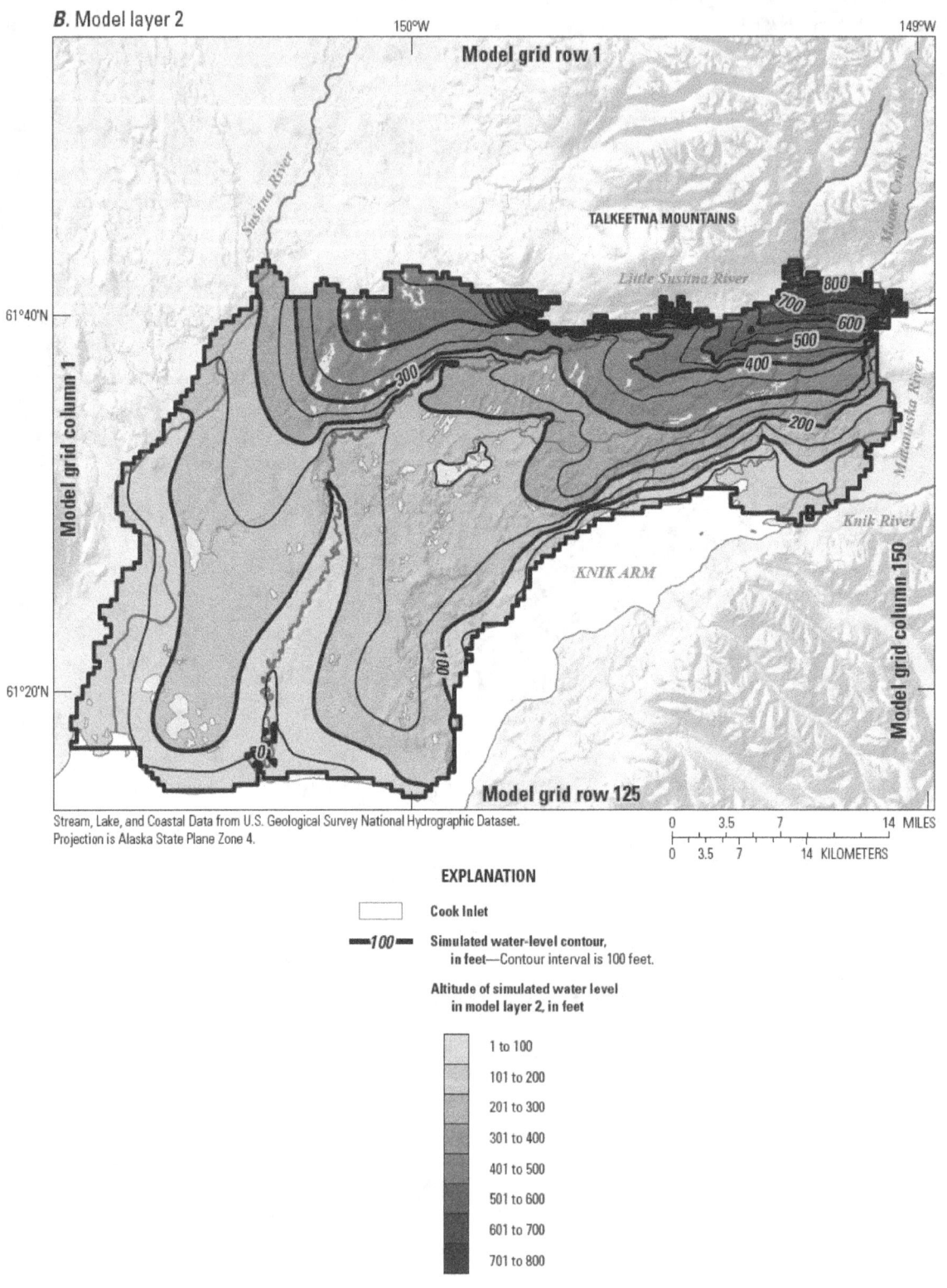

Figure 30.   Continued.

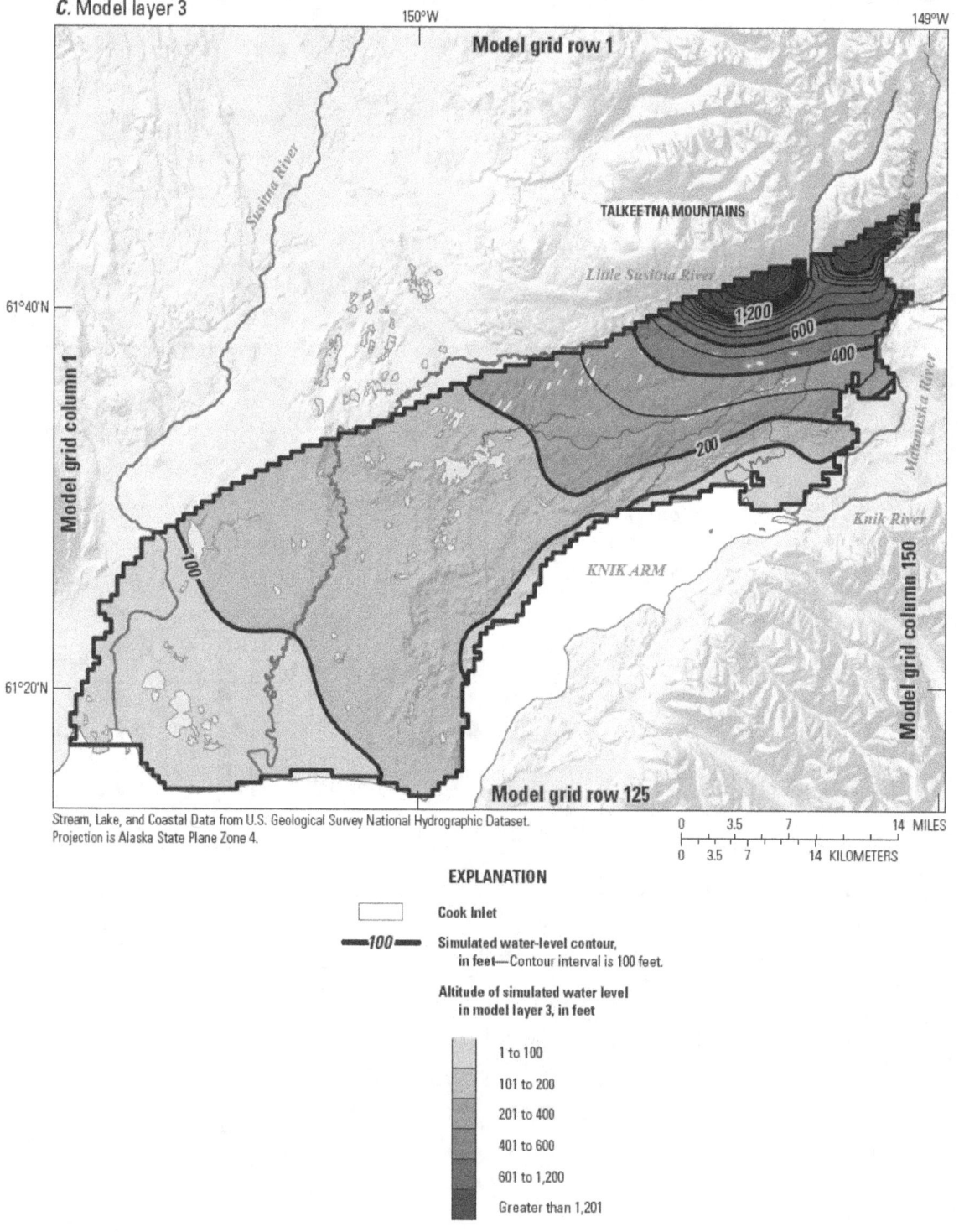

C. Model layer 3

EXPLANATION

Cook Inlet

—*100*— Simulated water-level contour,
in feet—Contour interval is 100 feet.

Altitude of simulated water level
in model layer 3, in feet

1 to 100

101 to 200

201 to 400

401 to 600

601 to 1,200

Greater than 1,201

Stream, Lake, and Coastal Data from U.S. Geological Survey National Hydrographic Dataset.
Projection is Alaska State Plane Zone 4.

Figure 30.   Continued.

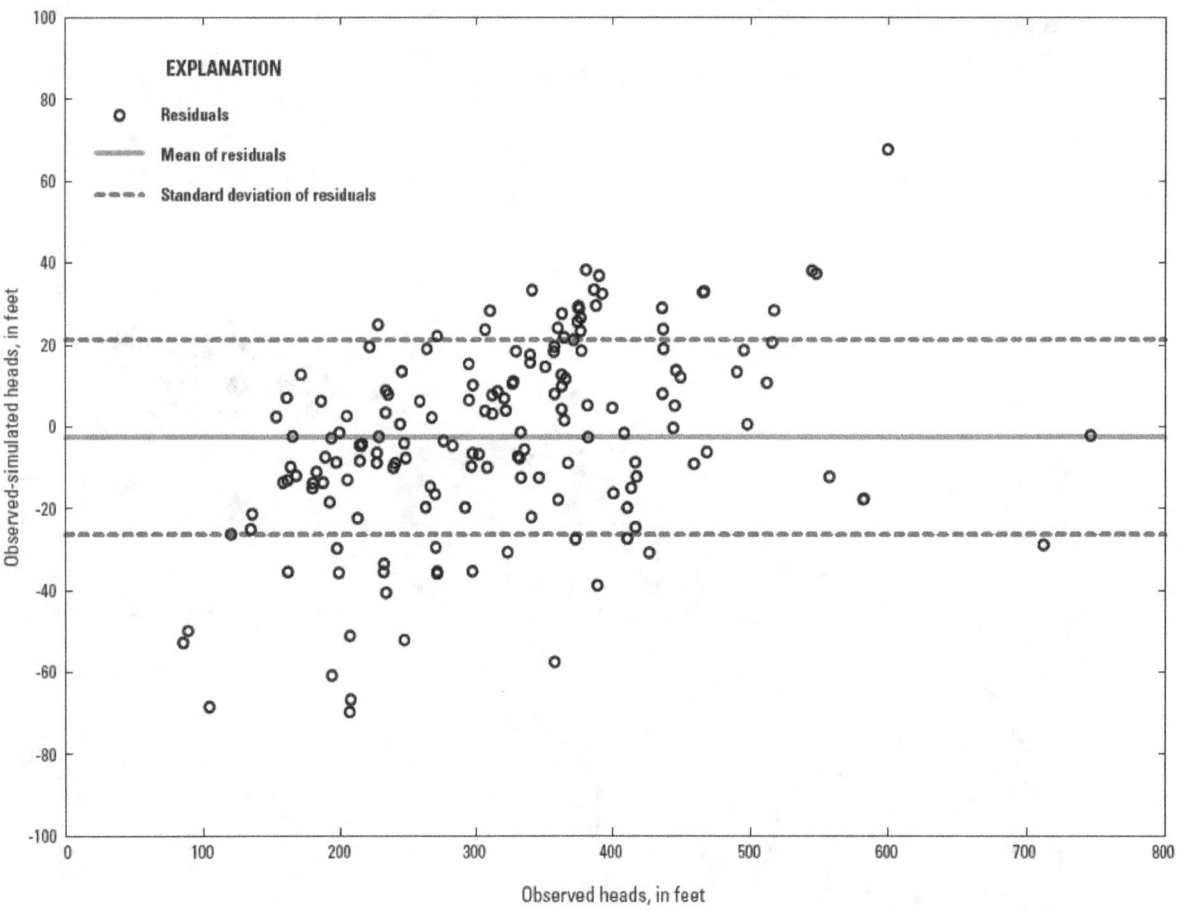

Figure 31.   Relation between hydraulic-head residuals and observed hydraulic heads, Matanuska-Susitna Valley, Alaska.

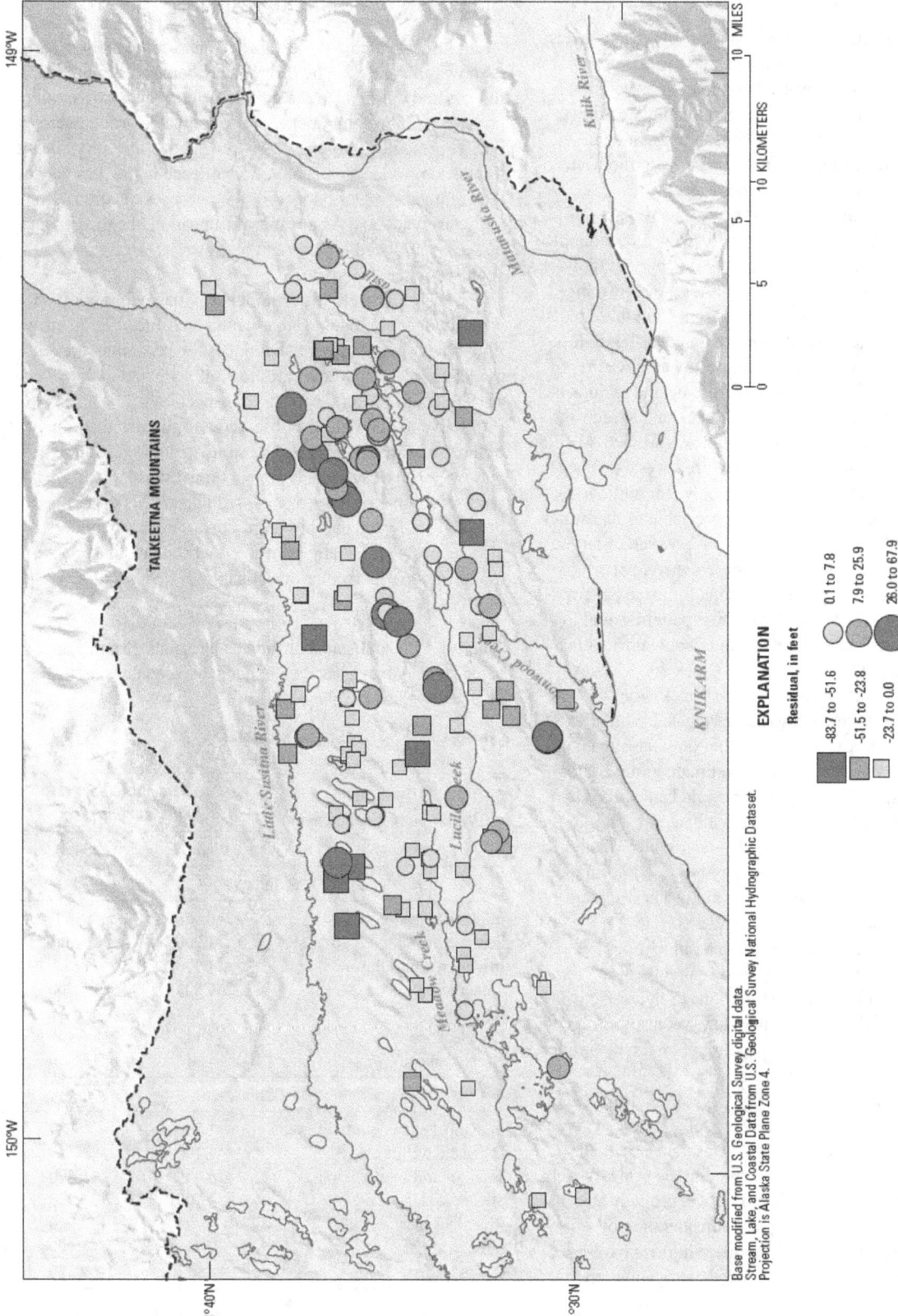

Base modified from U.S. Geological Survey digital data.
Stream, Lake, and Coastal Data from U.S. Geological Survey National Hydrographic Dataset.
Projection is Alaska State Plane Zone 4.

Figure 32. Spatial distribution of hydraulic-head residuals, Matanuska-Susitna Valley, Alaska. Negative values indicate that simulated heads are higher than observed heads; positive values indicate that simulated heads are lower than observed heads.

Simulated Flows

The flow residuals are defined as the difference between the observed and simulated flows; a positive value means that the model-simulated flows are less than observed flows, and a negative value means that the model-simulated flows are greater than observed flows. The mean and standard deviation of flow residuals calculated from the groundwater flow model are 0.4 and 2.8 ft^3/s, respectively. In general, the total simulated volumetric outflow rates to Wasilla Creek, Lucile Creek, and Meadow Creek, were close to, but consistently greater than, the measured volumetric outflow rates. Observed and simulated outflow rates are shown in figures 33A-C. Several patterns are apparent from these plots: first, the model-simulated outflow rates are consistently less than measured outflow rates in the most upstream reaches of the streams. Wasilla Creek originates from upland springs near the Little Susitna River Valley; Lucile Creek and Meadow Creek originate from lake-wetland complexes west of Wasilla. A second consistent pattern is that there are several reaches along all three creeks for which no groundwater outflow to streams was observed but for which appreciable groundwater outflow was computed. Much of the error between observed and simulated flows is attributed to these general patterns, warranting discussion of possible reasons for these patterns.

One possible explanation for the first pattern is that the modeled drain areas include the stream but not surrounding lakes or wetlands in the case of Lucile Creek and Meadow Creek and springs in the case of Wasilla Creek. For headwaters reaches where numerous spring-fed tributaries contribute water to streamflow, including only the stream as a drain may cause the model to underestimate groundwater discharge to the stream. This problem could be resolved by including the modeled drain areas in headwater reaches of the small streams considered in this study—especially Wasilla Creek and Meadow Creek. However, hydrologic information is insufficient to delineate a zone of groundwater discharge surrounding the headwaters of these streams.

The second pattern may be explained in part by heterogeneity of aquifer and streambed hydraulic properties. Stream reaches with identical distributions of hydraulic heads in the surrounding aquifer may receive vastly different quantities of groundwater inflow depending on the hydraulic conductivity of the surrounding aquifer and streambed material. The spatial resolution of hydrogeologic features represented in the groundwater flow model is coarse, and uniform drain conductance values were assigned over all reaches of each small stream represented in the groundwater flow model. The spatial discretization in the groundwater flow model is also coarse, preventing consideration of local-scale hydrogeologic detail. As a result, it is possible, and indeed likely, that some subsurface heterogeneity in the study area is not accurately represented by the groundwater flow model. The coarse resolution of the groundwater flow model may therefore lead to errors in the spatial distribution of model-simulated groundwater outflow along streams.

Simulated Groundwater Budget

The simulated groundwater budget for the modeled area is shown in table 15. The discrepancy between simulated inflows and outflows to the groundwater system was small (1 percent). The components of the simulated groundwater budget presented in table 15 are truncated at three significant figures to provide a measure of confidence in the accuracy of the volumetric water budget for the purposes of comparison with observed data. As a result of truncation, the discrepancy between simulated inflows and outflows reported in table 15 is slightly higher (5 percent).

The groundwater flow model provides information on water-budget components that would be difficult to quantify using direct measurement techniques. For example, the net model-simulated outflow rates to Knik Arm and major rivers are 9,320 and 159,000 acre-ft/yr, respectively. It is also instructive to compare the observed and simulated components of the groundwater budget. The simulated groundwater outflow to small streams is almost three times larger than the outflow volume estimated using field data. This discrepancy between simulated and observed flows across the modeled area suggests systematic error in either the field data or the model and merits further investigation.

Table 15.   Comparison of observed and simulated components of the steady-state groundwater budget components for the Matanuska-Susitna Valley, Alaska.

[**Abbreviations:** nd, no data; na, not applicable]

Water-budget component	Net volumetric flow rate (acre-feet per year)	
	Observed	**Simulated**
Inflows		
In-place recharge	260,000	na
Septic effluent	4,900	na
Irrigation return flows	1,900	na
Total recharge	266,800	267,000
River leakage	nd	0
Lake leakage	nd	0
Ocean leakage	nd	0
Outflows		
Municipal and community wells	1,700	na
Domestic wells	4,100	na
Total groundwater pumping	5,800	5,570
Stream leakage	10,700	31,900
River leakage	nd	159,000
Lake leakage	nd	46,900
Ocean leakage	nd	9,320

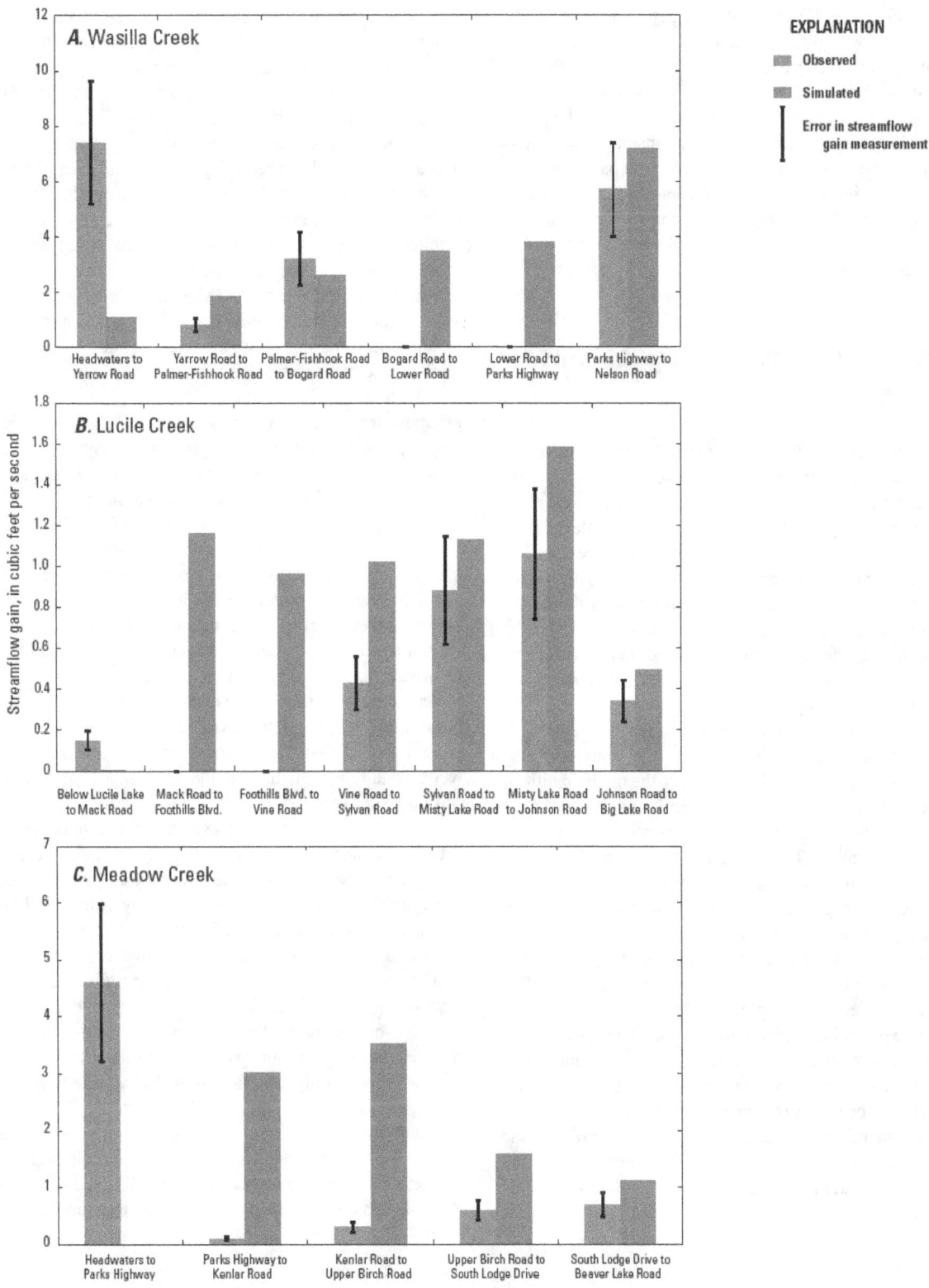

Figure 33. Observed and simulated streamflow gains. (*A*) Wasilla creek; (*B*) Lucile Creek; (*C*) Meadow Creek. Error bars on observed streamflow gains represent the standard deviation of observed gain.

MODFLOW-NWT automatically adjusts the specified flow rates for a well when the head in the model cell containing that well approaches the cell bottom. The threshold head value for initiating this adjustment procedure is controlled by an optional parameter, PHIRAMP, in the input for the Well Package. Unrealistic reductions in the specified flow rates for wells were minimized when the parameter PHIRAMP was set equal to 0.005. Nonetheless, there is a slight discrepancy (4 percent) between the specified pumping rates and the total withdrawals from wells in the simulated groundwater budget.

Model Limitations and Suggestions for Future Work

The groundwater flow model presented in this report provides a good fit to observed water levels and flows during 2009--11. Furthermore, the model reasonably represents important regional flow characteristics, including flow-path directions moving out of the Little Susitna Valley and the directions of water exchange with small streams. This model is also subject to limitations, and those limitations should be taken into consideration when using the model.

The scale of the model and the level of detail considered in spatial discretization and hydrogeologic characterization are appropriate for studying regional-scale groundwater availability issues. For example, this model could be used to simulate long-term hydrologic effects of groundwater withdrawals within the MSB core area or the municipalities of Palmer, Wasilla, or Houston, including changes in groundwater levels or changes in groundwater outflows to surface-water features. This capability of the model would be useful in determining if groundwater supplies are adequate for existing and proposed development in the study area that would require either the installation of new supply wells or increasing groundwater withdrawals from existing wells. This model could also be used to study long-term hydrologic effects of increases or decreases in groundwater recharge associated with change in land-use, land-cover, or climate variables. Such analysis would require modification of the DPM input files to recalculate groundwater recharge over the modeled area. This model can also be used to estimate net groundwater outflows to surface-water features, including rivers, small streams, and lakes. This capability would be useful in determining how, if at all, groundwater pumping affects in-stream flows in the modeled area.

This model is not appropriate for site-specific groundwater problems, including analysis of capture zones for individual wells. Also, contaminant-transport modeling applications require a far greater level of hydrogeologic detail than is formulated in this model. Therefore, this model is not suitable for contaminant-transport model applications. To address such problems requiring finer spatial discretization and hydrogeologic detail, it would be appropriate to use this regional-scale model to generate boundary conditions for smaller site-specific groundwater problems. Alternately, this model could be adapted to site-specific groundwater problems using Local Grid Refinement (Mehl and Hill, 2005).

Simulating the groundwater system as steady state assumes that aquifer inflows are equal to outflows, and there is no change in storage in the system. Using a steady-state model to simulate effects of stresses such as groundwater pumping assumes that such stresses propagate instantaneously throughout the groundwater system. The results of such analysis represent long-term effects of stresses, after equilibrium has been re-established. In reality, the groundwater system does not respond instantaneously to stresses; therefore, it is important to exercise caution when interpreting model results. This model was developed to simulate average long-term behavior of the groundwater system and cannot be used to simulate seasonal or interannual variability in the groundwater system. A transient model could be used to simulate time-varying response of the groundwater system to stresses such as seasonally varying recharge or groundwater pumping. Developing a reliable transient groundwater flow model for the Matanuska-Susitna Valley would require information on aquifer storage properties and continued collection of groundwater-level data.

Water-level contour maps synthesized from previous data and data collected during this study indicate directions of groundwater flow in relation to hydrologic boundaries such as large rivers and the ocean. However, hydrologic field data for the Susitna River or Little Susitna River are insufficient to accurately estimate the magnitude of water flows between groundwater and these boundaries. It is possible that long-term streamflow records for the Matanuska River could be used to estimate cumulative groundwater discharge to the river. However, the groundwater flow model presented here includes flow to only a portion of the Matanuska River. Extending the model domain up the Matanuska River Valley would allow for comparison of simulated and measured base flow in the Matanuska River. In addition, hydrologic field data characterizing groundwater discharge to Knik Arm would be valuable in assessing the groundwater budget simulated by the model.

Flow through the regional groundwater system is driven by recharge in the Little Susitna River Valley—including Archangel Creek and Fishhook Creek—Little Susitna River HUC-12 units. Much of this area, however, includes thin unconsolidated sediments on top of impermeable basement rock and is not represented by active cells in the groundwater flow model. The groundwater flow model presented in this report implicitly routes water falling on these relatively impermeable surfaces through the stage specified in the River Package. Using the Streamflow Routing Package (SFR) to represent the Little Susitna River would allow for more realistic description of overland flow processes in this area. However, the SFR package also requires more detailed information on channel geometry and roughness. In addition, future seepage-measurement campaigns on the Little Susitna River could be improved by measuring streamflow in tributaries to better isolate the exchange between groundwater and surface water.

The hydrogeologic framework model used in this study is based on an interpretation of unconsolidated sediments in the Meadow Lakes area as glacially deposited. However, it is also possible that these sediments were deposited by a paleoflood from a catastrophic outburst of glacial Lake Ahtna. Using this geochronology would likely alter the interpretation of regional hydrogeology, with implications for the distribution of aquifer hydraulic properties. Within both geochronologies, the borehole lithologic data available from water well driller's logs are prone to multiple interpretations. One way to account for uncertainty in the hydrogeologic framework model would be to construct alternate models to explain existing hydrologic data. This approach is particularly important in glacial aquifer systems such as the Matanuska-Susitna Valley aquifer system. Hydrogeologic conceptualization is also constrained by the lack of borehole data west of the core area. New borehole lithologic data will help to improve hydrogeologic understanding in this part of the study area.

Summary and Conclusions

The groundwater flow model presented in this report is one part of a comprehensive study of groundwater resources in the Matanuska-Susitna Valley, Alaska. This cooperative study between the U.S. Geological Survey and the Alaska Department of Natural Resources was initiated to provide information about regional-scale groundwater availability in the fastest growing part of Alaska. The two main objectives of this study were to perform a detailed characterization of hydrogeologic conditions in the aquifer system and to develop a computer model that could accurately simulate groundwater observations and be used in analysis of regional-scale groundwater-availability issues. This study built on previous studies of groundwater flow patterns, groundwater quality, and groundwater/surface-water interaction in the study area.

Existing subsurface lithologic data—primarily from well driller's logs—were compiled in a database. Lithologic data were used in conjunction with surficial geologic maps to construct two-dimensional hydrogeologic sections throughout the study area. During this process, five main hydrogeologic units were identified: Holocene Outwash and Alluvium, Naptowne Moraine, Fine Sediments, Lower Permeable Sediments, and Tertiary Sedimentary Rocks. A hydrogeologic framework model—including the spatial extent and thickness of each hydrogeologic unit—was then constructed using the two-dimensional hydrogeologic sections. The unconsolidated hydrogeologic units in general encompass heterogeneous facies corresponding to different depositional environments.

The isotopic composition of groundwater, surface-water, and precipitation samples collected in the study area indicates that groundwater comprises a mixture of waters recharged at relatively high elevations in the Talkeetna Mountains and waters recharged at lower elevations on the valley floor. Some spatial patterns in the isotopic composition of groundwater are apparent, notably isotopically depleted groundwater in the alluvial aquifer of the Matanuska River. Longitudinal variability in the isotopic composition of stream waters collected during this study, specifically in Lucile Creek, Wasilla Creek, and Meadow Creek, were attributed to the influence of upwelling groundwater. The results of the hydrogeologic framework model and analysis of water isotopic compositions indicate that groundwater and surface water are strongly interconnected in the Matanuska-Susitna Valley.

Components of the groundwater budget for the study area were defined using several different methods. Natural in-place recharge was estimated using a land-surface water-balance model. The magnitude of groundwater recharge varied over the course of the year; the bulk of groundwater recharge occurred during and after the late summer rainy season in July and August. The modeled in-place recharge rates are highest in the Little Susitna River Valley and Moose Creek River Valley. This finding is consistent with steep observed horizontal hydraulic gradients in these areas. Recharge to the groundwater system from septic effluent was estimated using information on the distribution of developed parcels outside municipal sewer service areas. The results of this analysis indicate that groundwater recharge rates from septic effluent are several orders of magnitude lower than recharge rates from natural in-place recharge.

Flow rates between groundwater and small streams were estimated during several field investigations, including seepage runs and analysis of synoptically collected stream-water and groundwater samples. These investigations indicated that Wasilla Creek, Meadow Creek, and Lucile Creek receive large amounts of groundwater discharge. As a result, it was important to include these features as boundaries in the numerical groundwater flow model. Water-use data from high-capacity wells are available between 2004 and

2010. Groundwater discharge to Knik Arm by seepage faces, springs, and diffuse submarine seepage and groundwater discharge to the Matanuska River, Little Susitna River, and Susitna River are likely dominant components of the regional groundwater budget. However, data were insufficient to quantify these groundwater outflows.

Long-term records from 1940 to 1993 show substantial natural fluctuation in groundwater levels. Historically, groundwater-level declines are typically observed following a succession of years with annual precipitation deficits of several inches, leading to below-average in-place recharge. Precipitation deficits of this magnitude were observed from 2008 to 2011. Observed water-level declines in wells across the study area are consistent with historical records of groundwater-level declines in response to below-average in-place recharge. Relatively short groundwater residence times in the permeable shallow aquifer units of the Matanuska-Susitna Valley, combined with abundant groundwater outflow to surface water, suggest that the groundwater system responds quickly to stresses and may be considered to be in a dynamic steady-state condition.

A steady-state groundwater model was developed to simulate long-term flow patterns in the regional aquifer system. The groundwater model encompasses approximately 1,200 mi^2 and is bounded by the Talkeetna Mountains to the north, Knik Arm to the south, the Matanuska River to the east, and the Susitna River to the west. The model grid consists of 3 layers, 125 rows, and 150 columns, with uniform cell area of 2,000 by 2,000 ft. The thickness and elevation of cell bottoms were defined to match the spatial extent and thickness of the five hydrogeologic units in the study area. In addition, model zones were defined in model layers 1 and 2 corresponding to the areal extent of different hydrogeologic units. The groundwater model simulates flow in the regional aquifer system subject to hydrologic stresses including recharge, groundwater withdrawals, and groundwater/surface-water interaction. Groundwater flow was simulated using stress packages from MODFLOW-NWT. The River Package was used to simulate interaction between groundwater and Moose Creek, the Matanuska River, the Little Susitna River, and the Susitna River. The Drain Package was used to simulate groundwater outflow to Wasilla Creek, Meadow Creek, and Lucile Creek. The General Head Boundary package was used to simulate interaction between groundwater and lakes. The groundwater flow equation was solved using MODFLOW-NWT. This approach improved model stability and reduced numerical error associated with the presence of extensive thin, unconfined aquifer units hydraulically connected to surface-water bodies.

During summer 2009, a synoptic water-level measurement was undertaken, including 135 measurements of groundwater levels and 38 measurements of lake stage. In addition, field investigations of groundwater/surface-water interaction along Wasilla Creek, Meadow Creek, and Lucile Creek yielded estimates of groundwater outflow to those streams. These water-level and flow data were used during the process of model calibration. The computer program UCODE-2005 was used to estimate the values of 13 model parameters, including horizontal hydraulic conductivity in the zones of model layers 1 and 2, riverbed vertical hydraulic conductivity in rivers, and the boundary conductance of small streams and lakes. The values of the model parameters obtained from the calibration process are within the range of values to be expected for unconsolidated sediments in the study area.

The simulated water-level contours from model layer 1 match the most prominent patterns in previously published water-table maps for the study area. The performance of the groundwater flow model was further assessed by analysis of hydraulic-head residuals and comparison of observed and simulated flows. Model calibration resulted in an even distribution of hydraulic-head residuals about the zero-line. Patterns were observed in the spatial distribution of hydraulic-head residuals surrounding a water-table ridge north of Wasilla, suggesting that model refinement is required to accurately simulate elevations of groundwater levels in this area. The model-simulated direction of exchange between groundwater and small streams matches observations. However, there is some error in the distribution of outflows along the length of each stream and in the cumulative outflow. In particular, the simulated outflows in the headwaters of Wasilla Creek and Meadow Creek are much less than observed outflows. This error is likely due in part to the limited spatial extent of the drain features assigned to represent those streams.

The scale of the groundwater flow model is appropriate for regional-scale analysis of groundwater availability and flow patterns but not for site-specific groundwater problems. However, the groundwater flow model presented in this report can be used to generate physically realistic boundary conditions for smaller-scale problems. The simulated hydraulic heads and flows from the groundwater model are representative of the long-term system response to hydrologic stresses. Analysis of changes in the groundwater system over time would require the development of a transient groundwater flow model, including estimation of aquifer storage properties. Continued monitoring of groundwater levels at stations established during this study will provide a basis for developing a transient flow model. The groundwater system in the Matanuska-Susitna Valley is complex because of the suite of depositional processes contributing to shallow aquifers and abundant groundwater/surface-water interaction. Simulation of this groundwater system could be improved by testing alternate hydrogeologic conceptualizations for the system and by continuing to quantify interaction between groundwater and surface water in the study area.

Acknowledgments

This work would not have been possible without the cooperation of hundreds of landowners who provided access to their property and water wells for water-level measurements, lake instrumentation, and groundwater-quality sampling. In addition, the availability of subsurface lithologic data from well driller's logs is a direct result of years of work by Roy Ireland of ADNR in creating and maintaining the WELTS database. Valuable field assistance was provided by Shamariah Hale (University of Alaska Fairbanks) and Charles Grammer (University of Alaska Anchorage). Brian Sweeney (University of Freiburg) assisted with collection of field data and entry of borehole lithology data during summer 2011. John Vaccaro of the USGS provided technical assistance with the use and modification of the Deep Percolation Model code for land-surface conditions in Alaska.

References Cited

Alaska Department of Administration, 2010, Alaska Oil and Gas Conservation Commission, accessed February 6, 2010, at http://doa.alaska.gov/ogc/.

Alaska Department of Natural Resources, 2009, Alaska Well Log Tracking System (WELTS): Division of Mining, Land and Water, Alaska Hydrologic Survey database, accessed January 2012 at http://www.navmaps.alaska.gov/welts/.

Brabets, T.P., Nelson, G.L., Dorava, J.M., and Milner, A.M., 1999, Water-quality assessment of the Cook Inlet Basin, Alaska—Environmental setting: U.S. Geological Survey Water-Resources Investigations Report 99-4025, 67 p.

Burns, L.E., 1985, The Border Ranges ultramafic and mafic complex, south-central Alaska—Cumulate fractionates of island-arc volcanics: Canadian Journal of Earth Science, v. 22, p. 1020–1038.

Clardy, B.I., Hanley, P.T., Hawley, C.C., and LaBelle, J.C., 1984, Guide to the bedrock and glacial geology of the Glenn Highway, Anchorage to the Matanuska Glacier and the Matanuska Coal Mining District: Anchorage, Alaska, Alaska Geological Society, Field Guide FG03, 63 p.

Clark, I.D., and Fritz, Peter, 1997, Environmental isotopes in hydrogeology: Boca Raton, Fla., CRC Press LLC, 352 p.

Craig, Harmon, 1961, Isotopic variations in meteoric waters: Science, v. 133, no. 3465, p. 1702–1703.

Driscoll, F.G., 1986, Groundwater and wells (2d ed.): St. Paul, Minn., Johnson Filtration Systems Inc., 1089 p.

Freethey, G.W., and Scully, D.R., 1980, Water resources of the Cook Inlet basin, Alaska: U.S. Geological Survey Hydrologic Investigations Atlas 620, 4 sheets, scale 1:1,000,000.

Freeze, R.A., and Cherry, J.A., 1979, Groundwater: Prentice-Hall, Englewood Cliffs, New Jersey, 604 p.

Glass, R.L., 2001, Ground-water quality in the Cook Inlet Basin, Alaska, 1999: U.S. Geological Survey Water-Resources Investigations Report 01-4208, 58 p.

Gracz, Mike, 2009. Matanuska-Susitna Wetland Mapping, Cook Inlet Wetlands, Alaska, accessed January 8, 2013 at http://cookinletwetlands.info/downloads/matsudownloads.htm.

Harbaugh, A.W., 2005, MODFLOW-2005, The U.S. Geological Survey modular ground-water model—The Ground-Water Flow Process: U.S. Geological Survey Techniques and Methods, book 6, chap. A16, variously paged.

Hill, M.C., 1998, Methods and guidelines for effective model calibration: U.S. Geological Survey Water-Resources Investigations Report 98-4005, 90 p.

Hill, M.C., and Tiedeman, C.R., 2007, Effective groundwater model calibration—With analysis of data, sensitivities, predictions, and uncertainty: Hoboken, New Jersey, John Wiley and Sons, 455 p.

International Atomic Energy Agency, 2009, Reference sheet for VSMOW2 and SLAP2 international measurement standards: Vienna, Austria, International Atomic Energy Agency, issued May 5, 2009, 5 p., accessed October 27, 2012, at http://nucleus.iaea.org/rpst/Documents/VSMOW2_SLAP2.pdf.

Jokela, J.B., Munter, J.A., and Evans, J.G., 1990, Ground-water resources of the Palmer-Big Lake area, Alaska—A conceptual model: Alaska Department of Natural Resources Division of Geological and Geophysical Surveys Report of Investigations 90-4, 38 p., 3 sheets, scale 1:25,000.

Kikuchi, C.P., Ferré, T.P.A., and Welker, J.M., 2012, Spatially telescoping measurements for improved characterization of groundwater—surface water interactions: Journal of Hydrology, v. 446–447, p. 1–12, accessed January 8, 2013 at http://dx.doi.org/10.1016/j.jhydrol.2012.04.002.

Kirschner, C.E., and Lyon, C.A., 1973, Stratigraphic and tectonic development of Cook Inlet Petroleum Province, in Pitcher, M.G., ed., Arctic geology—Proceedings of the Second International symposium on arctic geology: American Association of Petroleum Geologists Memoir 19, San Francisco, Calif., Feb. 1–4, 1971, p. 396–407.

Leake, S.A., and Claar, D.V., 1999, Procedures and computer programs for telescopic mesh refinement using MODFLOW: U.S. Geological Survey Open-File Report 99-238, 53 p.

Mathers, S.J., Wood, B., and Kessler, H., 2011, GSI3D 2011 software manual and methodology: British Geological Survey Internal Report, OR/11/020, 152 p.

Mehl, S.W., and Hill, M.C., 2005, MODFLOW-2005, The U.S. Geological Survey modular ground-water model— Documentation of shared node local grid refinement (LGR) and the Boundary Flow and Head (BFH) Package: U.S. Geological Survey Techniques and Methods, book 6, chap. A12, 68 p.

Moran, E.H., and Solin, G.L., 2006, Preliminary water-table map and water-quality data for part of the Matanuska-Susitna Valley, Alaska, 2005; U.S. Geological Survey Open-File Report 2006-1209, 43 p.

Munter, J.A., 2010, Potential hydrologic effects of gravel extraction south of Palmer, Alaska: J.A. Munter Consulting report for the Matanuska-Susitna Borough, 31 p.

National Oceanic and Atmospheric Administration, 2011, National Climatic Data Center: accessed December 2011, at http://lwf.ncdc.noaa.gov/oa/ncdc html.

Niswonger, R.G., Panday, Sorab, and Ibaraki, Motomu, 2011, MODFLOW-NWT, A Newton formulation for MODFLOW-2005: U.S. Geological Survey Techniques and Methods, book 6, chap. A37, 44 p.

Poeter, E.P., Hill, M.C., Banta, E.R., Mehl, Steffan, and Christensen, Steen, 2005, UCODE_2005 and six other computer codes for universal sensitivity analysis, calibration, and uncertainty evaluation: U.S. Geological Survey Techniques and Methods, book 6, chap. A11, 283 p.

Razack, M., and Huntley, D., 1991, Assessing transmissivity from specific capacity in a large and heterogeneous alluvial aquifer: Ground Water, v. 29, no. 6, p. 856-861.

Reger, R.D., and Updike, R.G., 1983, Upper Cook Inlet region and the Matanuska Valley, in Pewe, T.L., and Reger, R.D., eds., Guidebook to permafrost and Quaternary geology along the Richardson and Glenn Highways between Fairbanks and Anchorage, Alaska: Fourth International Conference on Permafrost, Fairbanks, July 18-22, 1983, Field Trip Guidebook 1—Alaska Division of Geological and Geophysical Surveys, Fairbanks, Alaska, p. 185–259.

Rutledge, A.T., 1998, Computer programs for describing the recession of ground-water discharge and for estimating mean ground-water recharge and discharge from streamflow records—Update: U.S. Geological Survey Water-Resources Investigations Report 98-4148, 43 p.

Shellenbaum, D.P., Gregersen, L.S., and Delaney, P.R., 2010, Top Mesozoic unconformity depth map of the Cook Inlet Basin, Alaska: Division of Geological and Geophysical Surveys Report of Investigations 2010-2, 1 sheet, scale 1:500:000.

Trainer, F.W., 1960, Geology and ground-water resources of the Matanuska Valley agricultural area, Alaska: U.S. Geological Survey Water-Supply Paper 1494, 116 p., 3 plates.

Trop, J.M., and Ridgway, K.D., 1999, Sedimentology and Provenance of the Paleocene-Eocene Arkose Ridge Formation, Cook Inlet-Matanuska Valley Forearc Basin, Southern Alaska, in Pinney, D.S., and Davis, P.K., eds., 2000, Short Notes on Alaska Geology 1999: Alaska Division of Geological and Geophysical Surveys Professional Report 119, 152 p.

Tucci, P., 1982, Use of a three-dimensional model for analysis of the ground-water flow system in the Parker Valley, Arizona and California: U.S. Geological Survey Open-File Report 82-1006, 54 p.

University of Alaska Anchorage, 2012, Alaska water isotopes network: Environment and Natural Resources Institute database, accessed January 30, 2012, at http://www.uaa.alaska.edu/enri/databases/AKWIN/DataSets/index.cfm.

U.S. Department of Agriculture, Natural Resources Conservation Service, 2011a, Alaska SNOTEL sites: Anchorage, Alaska, National Resources Conservation Service database, accessed September 1, 2011, at http://www.wcc nrcs.usda.gov/snotel/Alaska/alaska html.

U.S. Department of Agriculture, Natural Resources Conservation Service, 2011b, Alaska soil survey information: Palmer, Alaska, National Resources Conservation Service database, accessed October 1, 2011, at http://www.ak nrcs.usda.gov/soils/index.html.

U.S. Department of Agriculture, Natural Resources Conservation Service, 2011c, Geospatial data gateway: National Resources Conservation Service database, accessed October 1, 2011, at http://datagateway nrcs.usda.gov/GDGHome.aspx.

U.S. Department of Agriculture, Natural Resources Conservation Service, 2011d, U.S. general soil map (STATSGO2): National Resources Conservation Service database, accessed January 8, 2013, at http://soils.usda.gov/survey/geography/ssurgo/description statsgo2 html.

U.S. Geological Survey, 2009, National Hydrography Dataset: U.S. Geological Survey database, accessed September 1, 2009, at http://nhd.usgs.gov/data.html.

U.S. Geological Survey, 2011a, National Elevation Dataset: U.S. Geological Survey database, accessed October 1, 2011, at http://ned.usgs.gov/.

U.S. Geological Survey, 2011b, The National Map Viewer and Download Platform: U.S. Geological Survey database, accessed September 1, 2011, at http://nationalmap.gov/viewer.html.

Usibelli Coal Mine, Inc., 2009, Surface Mining and Reclamation Control Act permit application, Part C: accessed January 8, 2013, at http://dnr.alaska.gov/mlw/mining/coal/wishbone/.

Vaccaro, J.J., 2007, A deep percolation model for estimating ground-water recharge—Documentation of modules for the modular modeling system of the U.S. Geological Survey: U.S. Geological Survey Scientific Investigations Report 2006-5318, 30 p.

Vaccaro, J.J., and Maloy, K.J., 2006, A thermal profile method to identify potential ground-water discharge areas and preferred salmonid habitats for long river reaches: U.S. Geological Survey Scientific Investigations Report 2006-5136, 16 p.

Wiedmer, M., Montgomery, D.R., Gillespie, A.R., and Greenberg, H., 2010, Late Quaternary megafloods from Glacial Lake Atna, southcentral Alaska, U.S.A.: Quaternary Research, v. 73, p. 413-424.

Wilson, F.H., Hults, C.P., Schmoll, H.R., Haeussler, P.J., Schmidt, J.M., Yehle, L.A., and Labay, K.A., comps., 2009, Preliminary geologic map of the Cook Inlet Region, Alaska—Including parts of the Talkeetna, Talkeetna Mountains, Tyonek, Anchorage, Lake Clark, Kenai, Seward, Iliamna, Seldovia, Mount Katmai, and Afognak 1:250,000-scale quadrangles: U.S. Geological Survey Open-File Report 09-1108, accessed January 8, 2013, at http://pubs.usgs.gov/of/2009/1108/.

Winter, T.C., Harvey, J.W., Franke, O.L., and Alley, W.M., 1998, Groundwater and surface water—A single resource: U.S. Geological Survey Circular 1139, 79 p.

Appendixes

Appendix tables are presented as Microsoft© Excel spreadsheets. They can be accessed and downloaded at http://pubs.usgs.gov/sir/2013/5049/.

Appendix A. Borehole Lithologic Data for Wells Used in Hydrogeologic Framework Model, Matanuska-Susitna Valley, Alaska.

Appendix B. Identification, Description, and References for Water Isotope Data in the Matanuska-Susitna Valley, 1999–2011.

Appendix C. Physical Characteristics of Hydrologic Response Units (HRU), and HRU Water Budgets Calculated in the Deep Percolation Model (DPM), Matanuska-Susitna Valley, Alaska.

Appendix D. Physical Data for Wells Included in the 2009 Synoptic Water Level Measurement, Matanuska-Susitna Valley, Alaska.